# Urological Cancer 2023-2025

Collection Editors


**José I. López**
**Claudia Manini**



MDPI

Basel • Beijing • Wuhan • Barcelona • Belgrade • Novi Sad • Cluj • Manchester

*Collection Editors*

José I. López
Biomarkers in Cancer Unit
Biobizkaia Health Research
Institute
Barakaldo
Spain

Claudia Manini
Department of Pathology
San Giovanni Bosco Hospital
Turin
Italy

*Editorial Office*
MDPI AG
Grosspeteranlage 5
4052 Basel, Switzerland

This is a reprint of the Topical Collection, published open access by the journal *Cancers* (ISSN 2072-6694), freely accessible at: https://www.mdpi.com/journal/cancers/topical_collections/C0M0D9M43H.

For citation purposes, cite each article independently as indicated on the article page online and as indicated below:

Lastname, A.A.; Lastname, B.B. Article Title. *Journal Name* **Year**, *Volume Number*, Page Range.

**ISBN 978-3-7258-4949-9 (Hbk)**
**ISBN 978-3-7258-4950-5 (PDF)**
**https://doi.org/10.3390/books978-3-7258-4950-5**

Cover image courtesy of José I. Lopez

# Contents

# About the Editors

**José I. López**

José I. López is an Advisor Researcher of the Biomarker in Cancer Unit at the Biocruces-Bizkaia Health Research Institute. He graduated from the Faculty of Medicine, University of the Basque Country, Leioa, Spain, and trained in pathology at the Hospital Universitario 12 de Octubre, Madrid, Spain. He received his PhD at the Universidad Complutense of Madrid, Spain. Dr. López has served as a pathologist for more than 30 years in several hospitals in Spain and subspecializes in uropathology, on which he has published more than 200 peer-reviewed articles and reviews. Dr. López is interested in translational uropathology in general and in renal cancer in particular and collaborates with several international research groups unveiling the genomic landscape of urological cancer. Intratumor heterogeneity, tumor sampling, the tumor microenvironment, tumor ecology, immunotherapy, and the basic mechanisms of carcinogenesis are his main topics of interest.

**Claudia Manini**

Claudia Manini is Head of the Department of Pathology at the San Giovanni Bosco Hospital in Turin, Italy. She graduated from the Faculty of Medicine and Surgery and completed a post-graduate in surgical pathology from the University of Turin, Turin, Italy. Dr. Manini has served as a pathologist for more than 25 years in several hospitals in Italy, developing expertise in diagnostic uropathology, neuropathology and gyneco-pathology. Her main interest is translational pathology.

cancers

MDPI

Article

# Gene-Transcript Expression in Urine Supernatant and Urine Cell-Sediment Are Different but Equally Useful for Detecting Prostate Cancer

**Marcelino Yazbek Hanna [1,†], Mark Winterbone [2,†], Shea P. O'Connell [2,†], Mireia Olivan [3,4], Rachel Hurst [2], Rob Mills [5], Colin S. Cooper [2], Daniel S. Brewer [2,6,*] and Jeremy Clark [2]**

1 Urology Department Castle Hill, Hull University Teaching Hospital, Castle Rd, Cottingham HU16 5JQ, UK
2 Norwich Medical School, University of East Anglia, Norwich NR4 7TJ, UK
3 Translational Oncology Laboratory, Department of Pathology and Experimental Therapy, School of Medicine, Universitat de Barcelona, 08907 L'Hospitalet de Llobregat, Spain
4 Program of Molecular Mechanisms and Experimental Therapy in Oncology (ONCOBELL), Bellvitge Biomedical Research Institute (IDIBELL), 08908 L'Hospitalet de Llobregat, Spain
5 Norfolk and Norwich University Hospitals NHS Foundation Trust, Norwich NR4 7UY, UK
6 Earlham Institute, Norwich NR4 7UZ, UK
* Correspondence: d.brewer@uea.ac.uk
† These authors contributed equally to this work.

**Simple Summary:** Cancer cells and vesicles are transported in prostatic secretions to the urethra and are flushed out on urination. These cells and vesicles contain prostate-specific gene transcripts, but their relative usefulness in prostate cancer detection has not been fully determined. We have examined the expression of 167 gene-probes in vesicle and cell fractions from 76 urine samples provided by men with and without prostate cancer. Measured gene expression profiles varied between the fractions. Many genes were useful as biomarkers for PCa in one fraction only, supporting the analysis of fractionated urine over the analysis of whole urine. Signatures constructed from cell or vesicle data were equally good at distinguishing prostate cancer from no-cancer controls. A combined-fraction signature did not show significant improvement. We present data on the relative expression of six housekeeping genes and the potential tissue origin of cells and vesicles in urine.

**Abstract:** There is considerable interest in urine as a non-invasive liquid biopsy to detect prostate cancer (PCa). PCa-specific transcripts such as the *TMPRSS2:ERG* fusion gene can be found in both urine extracellular vesicles (EVs) and urine cell-sediment (Cell) but the relative usefulness of these and other genes in each fraction in PCa detection has not been fully elucidated. Urine samples from 76 men (PCa $n = 40$, non-cancer $n = 36$) were analysed by NanoString for 154 PCa-associated genes-probes, 11 tissue-specific, and six housekeeping. Comparison to qRT-PCR data for four genes (*PCA3*, *OR51E2*, *FOLH1*, and *RPLP2*) was strong ($r = 0.51$–$0.95$, Spearman $p < 0.00001$). Comparing EV to Cells, differential gene expression analysis found 57 gene-probes significantly more highly expressed in 100 ng of amplified cDNA products from the EV fraction, and 26 in Cells ($p < 0.05$; edgeR). Expression levels of prostate-specific genes (*KLK2*, *KLK3*) measured were ~20× higher in EVs, while PTPRC (white-blood Cells) was ~1000× higher in Cells. Boruta analysis identified 11 gene-probes as useful in detecting PCa: two were useful in both fractions (*PCA3*, *HOXC6*), five in EVs alone (*GJB1*, *RPS10*, *TMPRSS2:ERG*, *ERG*_Exons_4-5, *HPN*) and four from Cell (*ERG*_Exons_6-7, *OR51E2*, *SPINK1*, *IMPDH2*), suggesting that it is beneficial to fractionate whole urine prior to analysis. The five housekeeping genes were not significantly differentially expressed between PCa and non-cancer samples. Expression signatures from Cell, EV and combined data did not show evidence for one fraction providing superior information over the other.

**Keywords:** urine; prostate; cancer; biomarker; extracellular vesicles; cell-sediment

## 1. Introduction

Prostate cancer (PCa) is the second most commonly diagnosed cancer in men in the world [1]. Suspicion of PCa is based on serum PSA testing, abnormal digital rectal examination (DRE) and more recently MRI [2]. Confirmation of PCa is by needle biopsy, an invasive procedure that can have significant morbidity. Improving the pre-screening methods used to decide who to biopsy would reduce costs and patient stress. The use of 'liquid biopsy' tests using samples of blood, saliva and urine have been investigated. Blood is used to examine PSA levels and also to detect circulating tumor cells [3] and cell-free nucleic acids, though dilution of markers in the large volume of circulating blood has made sensitivity an issue [4]. Saliva has been utilised to examine germline changes such as faulty DNA-repair genes that could result in a predisposition for cancer development. Urine in comparison is used to examine the presence or absence of prostate cancer within a prostate, and the connection of the prostate to the urinary system presents several advantages in PCa detection. The prostate is a secretory organ that drains into the urethra. Prostate cancers shed cells and extracellular vesicles (EVs) which are carried with these secretions and are flushed out of the body on urination in the first 15 mL of urine [5,6]. Urine, therefore, represents an attractive non-invasive liquid-biopsy source of PCa-biomarkers.

Urine studies have largely focussed on urine cell-sediment in which mRNA transcripts such as *PCA3* [7] and *TMPRSS2:ERG* [8] have shown diagnostic utility. The use of cell-free RNA harvested from EVs in urine supernatant is a promising alternative. EVs contain PCa-specific transcripts and EV membranes have been shown to protect from nucleases and other potentially harmful chemicals present in urine [9]. We have recently used EV expression data for 38 gene-probes to construct Prostate Urine Risk (PUR) signatures, which have shown the potential to predict disease progression over a five-year follow-up period (HR = 8.2) [10]. Only a few studies to date have attempted to compare cell-sediment and EV urine fractions, each only examining a handful of genes, with no consensus on each fraction's potential to differentiate between PCa and non-PCa. Dijkstra et al. [11] compared levels of *PCA3* and *TMPRSS2-ERG* in the cell-sediments and EVs from 30 men scheduled for a biopsy. 10% of cell-sediments were unusable due to the formation of crystals following centrifugation whilst none were lost in the EV fraction. Dijkstra recorded higher yields of RNA from cell-sediments than EVs and when using *PCA3* mRNA levels better diagnostic utility in the cell-sediments fraction was observed. In contrast, Pellegrini et al. [12] found that the EV fraction had higher total RNA yields ($n = 105$), better RNA quality as assessed by RIN-score and higher levels of *PCA3* and *ERG* RNAs ($n = 52$). Hendriks and colleagues [13] compared the expression of mRNA transcripts in whole urine, cell-sediment and EVs ($n = 29$). They observed that expression of *KLK3*, *PCA3* and *ERG* were highest in whole urine, followed by EV, and lowest in cell-sediments. They reported that *PCA3* transcripts were expressed significantly more highly in PCa patients compared to non-PCa in both the whole-urine and cell fractions but not in the EV fractions, while *ERG* was only significantly differentially expressed in the cell-sediment fraction. These studies suggest that, although urine EVs may provide a more robust source of biomarkers, the cell-sediment fraction appears to have greater diagnostic utility, albeit in only four gene transcripts examined.

Could combining the examination of transcripts in both Cell-sediment (Cell) and EVs improve the utility of urine biomarkers to detect prostate cancer? To examine this question, NanoString data from 167 gene-probes were interrogated (including the 38 PUR signature gene-probes) in Cell and EV fractions from 76 samples and correlated with PCa disease status on biopsy.

## 2. Materials and Methods

### 2.1. Clinical Cohort

Post-DRE urine samples were collected from 90 men attending the Norfolk and Norwich University Hospitals. Ethical approval for the study was gained from the East of England Research Ethics Committee, UK (ref 12/EE/0058). Men were divided by PCa status: Men with PCa on 10-core trans-rectal ultrasound-guided (TRUS) biopsy, and 'Non-

Cancer' (NC), which consisted of 7 unbiopsied men with a normal PSA for age [14] and 29 men with a raised serum PSA ($\geq$4 ng/mL) that were found to be negative for cancer on TRUS biopsy (see Table 1 for cohort clinical characteristics).

**Table 1.** Cohort characteristics. GG: Gleason Grade Group, No Bx: No Biopsy performed, Neg Bx: PCa-negative on biopsy, Age: median age in years (Y), IQR: interquartile range, N: Number of samples.

| Characteristic | Non-Cancer | Prostate Cancer |
|---|---|---|
| Number of Samples | 36 | 40 |
| Age (IQR) | 66 (12.3) | 70.0 (9.5) |
| PSA (ng/mL) (IQR) | 6.3 (4.0) | 9.1 (5.4) |
| Biopsy results (N, %) | No Bx (7, 19%)<br>Neg Bx (29, 79%) | GG1 (6, 26%)<br>GG2 (17, 74%)<br>GG3 (6, 35 %)<br>GG $\geq$ 4 (11, 65%) |

## 2.2. Cell and EV Fractions

Urine samples were centrifuged to pellet the cells, EVs were harvested from the supernatant by filtration using a 100 kDa spin-filter unit (Millipore, Burlington, MA, USA), and RNA was extracted from each fraction using RNeasy Micro columns (Qiagen #74004, Hilden, Germany) as in Connell et al. 2019 [10].

## 2.3. NanoString Data: Feature Selection and Analysis

Cell and EV RNA samples (5–20 ng) were amplified with the Ovation PicoSL WTA System V2 kit (Nugen #3312-48, Leeds, UK) and NanoString analysis for 167 gene-probes was performed on 100 ng amplified cDNA products at the Human Dendritic Cell Laboratory, Newcastle University, UK as described by Connell et al., [10] (see Table S1 for the 167 NanoString gene-probes used). Where multiple probes per transcript were used, the exact exons targeted are stated, e.g., *ERG_Exons_4-5* and *ERG_Exons_6-7*. NanoString data were subject to quality control prior to normalisation as per NanoString's guidelines. Fourteen samples were removed due to NanoString normalisation factors being outside the manufacturer's acceptable range (less than 0.1 or greater than 10.0, https://nanostring.com/support-documents/gene-expression-data-analysis-guidelines/ (accessed on 1 January 2022)). The measured expression levels are the counts for that probe detected in 100 ng of amplified cDNA products and should be considered as a proportion rather than the absolute total amount of RNA for that gene in a particular urine fraction.

All analyses were performed in R version 3.4.1. The *edgeR* package was used to pre-process and examine differential gene expression within and between the Cell and EV urine fractions (data in Table S2). *EdgeR* pre-processing implements the filtering strategy described by Chen et al. (2016) [15] and which retained genes that had a minimum of 10 counts in 5 samples, leaving 105/167 gene-probes for subsequent analysis. Biological variation across gene-probes was estimated based on the use of negative binomial distribution and generalised linear models [16,17].

Gene-probes useful in identifying PCa were selected by comparing Non-Cancer samples to PCa-samples. A robust feature selection workflow was implemented that used the Boruta algorithm [16] and bootstrap resampling as described in Connell et al. (2020) [17]. Boruta uses a random forest algorithm iteratively compared feature importance against randomly shuffled predictors named "shadow features". Features that performed significantly worse compared to the best-performing shadow feature (Shadow Max) at each permutation were consecutively dropped until only stable features remained. Gene-probes were identified as 'tentative' or 'confirmed' by comparison to the performance of the shadow features. 'Confirmed' indicates that a gene probe performed statistically better than the maximum performance of the shadow feature (ShadowMax)—this is a high threshold as the ShadowMax could easily be quite high by chance. 'Tentative' indicates that a gene probe performed significantly better than the mean performance of the Shadow gene-probe (ShadowMean) but was not statistically better than the ShadowMax.

The Boruta-selected gene-probes were combined in a random forest model to produce three risk scores for prostate cancer using data from (i) the Cell-sediment fraction, (ii) the EV fraction, and (iii) a combined risk score using data from both fractions.

### *2.4. RT-PCR Analysis*

Quantitative RT-PCR (qRT-PCR) was performed for *PCA3, OR51E2, FOLH1, KLK3,* and *RPLP2* following the protocol of Sequeiros et al. [18] (see Table S3 for PCR primers). qRT-PCR analysis of the samples was performed on a 384-well plate. Duplicate qRT-PCRs were run on a separate 384-well plate on the same day. The presence of *TMPRSS2:ERG* transcripts was detected using *TMPRSS2* exon1 and *ERG* exon 6 primers as in Clark et al. [19]. For comparison of qRT-PCR data with NanoString data for the same genes, we used Cohen's (1988) conventions to interpret effect size; small/weak: $r \geq 0.1$, moderate: $r \geq 0.3$, large/strong: $r \geq 0.5$ [20].

## 3. Results

### *3.1. Gene-Transcript Expression in Urine EV and Urine Cell-Sediment Are Different*

#### 3.1.1. No Differences Were Observed in the Expression of Housekeeper Genes between Cancer and Non-Cancer in Both Urine Fractions

We analysed Cell and EV fraction samples from urine using a 167-probe custom-built NanoString assay which contained six housekeeping gene-probes, five tissue-specific genes and 154 PCa-associated gene-probes (Table S1). No significant differences in housekeeper expression (*ALAS1, B2M, GAPDH, HPRT, RPLP2, TBP*) were found between non-cancer (NC) and prostate cancer (PCa) samples in either fraction (False Discovery Rate (FDR) $p < 0.05$; edgeR; Figure 1B,D; Table S2).

#### 3.1.2. EV and Cell Fractions Have a Different Profile of Tissue of Origin

To investigate the origin of the Cells and EVs found in urine, we analysed five tissue-specific gene-probes corresponding to the following tissue/cell types: normal prostate (*KLK2, KLK3*), bladder (*UPK2*), kidney (*SLC12A1*) and blood leukocytes (*PTPRC*) (Figure 1A,C). Median *KLK2* and *KLK3* expression levels measured in 100 ng of amplified cDNA products were ~20-fold higher in EV compared to the Cell fraction. *PTPRC*, a gene expressed in all nucleated cells of hematopoietic origin, was detected at a high value in the Cell fraction with only minimal levels of expression in the EVs (median > 1000-fold lower). Measured levels of bladder-specific *UPK2* and kidney-specific *SLC12A1* were low in both fractions. Differences between Non-Cancer (NC, $n = 36$) and PCa ($n = 40$) were only significant for *SLC12A1* in EVs (median levels ~50-fold higher in the PCa EV samples, FDR $p = 0.034$; edgeR; Table S2).

#### 3.1.3. Most Gene-Probes Are Differentially Expressed between EV and Cell Fractions

After pre-filtering for present probes ($n = 105$; Table S1), 83% of gene-probes were significantly differentially expressed in 100 ng of amplified cDNA products between the EV and Cell fractions (FDR $p < 0.05$; edgeR; Tabels 2 and S2). Of these, 57 were found to be significantly more highly expressed in the EV fraction and 26 were more highly expressed in the Cell fraction.

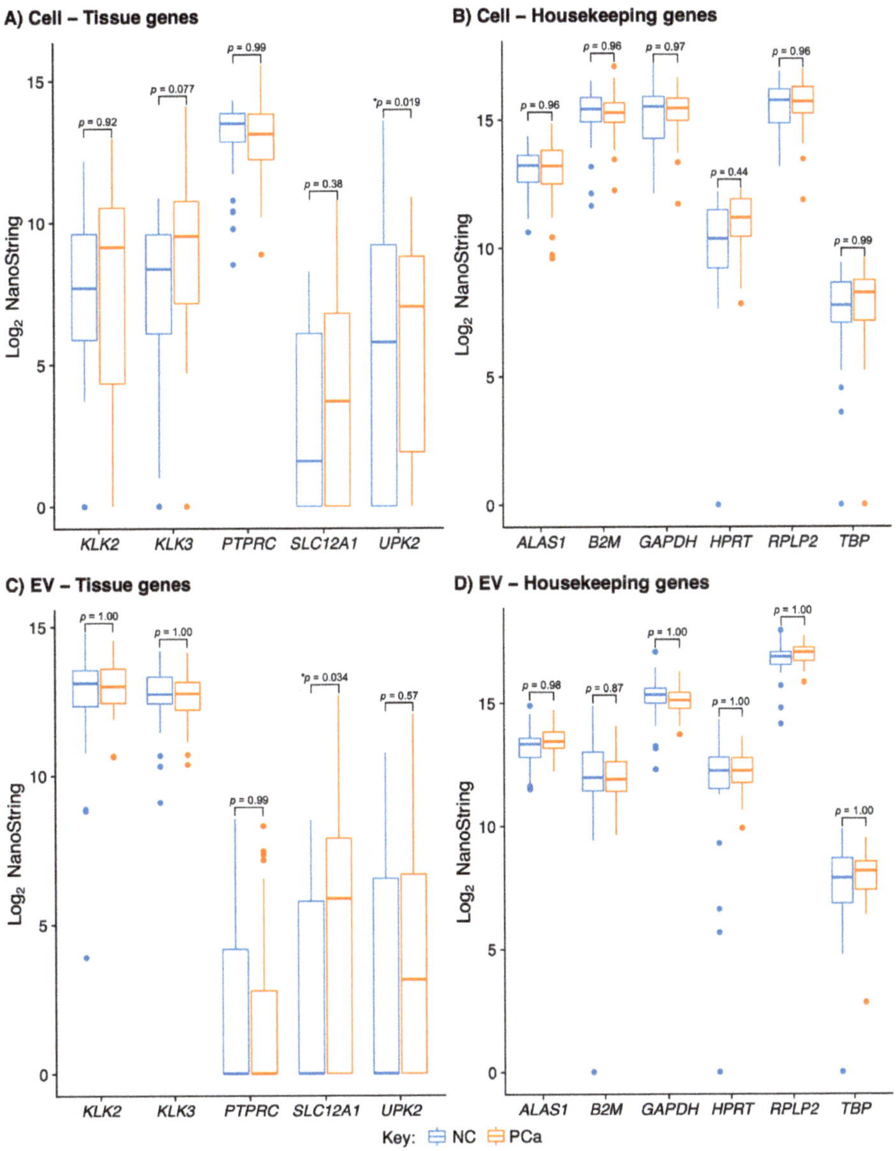


**Figure 1.** NanoString signals for 5 tissue-specific gene-probes (**A**,**C**) and 6 housekeeping gene-probes (**B**,**D**) in urine cell-sediment (Cell, upper panel) and urine extracellular vesicles (EV, lower panel). Blue indicates non-cancer samples (NC), and red indicates prostate cancer samples (PCa). $p$-values are for statistical difference between cancer and non-cancer by edgeR; * indicates a significant difference (False Discovery Rate $p < 0.05$).

**Table 2.** The top 10 differentially expressed genes between EV and Cell fractions. Log FC: log2 fold change, *p* value: False discovery rate significance (edgeR). Summary information on gene function and published information linked to PCa is provided here, and more detailed information on each gene is provided in Table S1.

| Gene Name | $Log_2$ FC | *p*-Value | Expression | Gene Function/Link to PCa |
|---|---|---|---|---|
| *NEAT1* | −8.65 | <0.00001 | Lower in EVs | Bone metastasis [21] |
| *MIR4435.1HG lOC541471* | −4.43 | <0.00001 | Lower in EVs | No publications on PCa |
| *IFT57* | 2.81 | <0.00001 | Higher in EVs | Pro-apoptotic [22] |
| *B2M* | −3.9 | <0.00001 | Lower in EVs | Housekeeper [23] |
| *BTG2* | −3.67 | <0.00001 | Lower in EVs | Tumor-suppressor [24] |
| *MCTP1* | −9.42 | <0.00001 | Lower in EVs | Calcium signaling [25] |
| *DPP4* | 3.05 | <0.00001 | Higher in EVs | Overexpressed in PCa [26] |
| *APOC1* | −8.16 | <0.00001 | Lower in EVs | Overexpressed in PCa [27] |
| *H1.2* | 1.68 | <0.00001 | Higher in EVs | Apoptotic response to DNA damage [28] |
| *ECI2* | 2.09 | <0.00001 | Higher in EVs | Knock-out may have a therapeutic response in PCa [29] |

#### 3.1.4. Expression Changes between Non-Cancer and Cancer Are Different in the EV and Cell Fractions

Thirteen probes/fraction combinations were significantly differentially expressed between NC and PCa in the Cell and EV expression data (FDR $p < 0.05$; edgeR; Table 3; data presented as a Volcano plot in Figure 2A)—7 in EVs and 6 in Cell; all were overexpressed in cancer apart from *CDKN3*, *SPINK1* and *UPK2*. Three commonly used urine biomarker genes were in the top 10 differentially expressed gene/fraction combinations: *ERG*, *HOXC6*, and *PCA3*. *ERG* and *HOXC6* were significantly more highly expressed in PCa in both Cell and EVs fractions, while *PCA3* was only significantly higher in PCa in EVs. Median expression levels of these three genes were higher in EV than Cell fractions (EV vs. Cell: *ERG* 46 vs. 0.5; *HOXC6* 1432 vs. 7.5; *PCA3* 2750 vs. 163).

**Table 3.** The significantly differentially expressed genes between non-Cancer and prostate cancer (PCa) in univariate analysis. Log FC: $log^2$ fold change, *p* value: False discovery rate significance (edgeR). Summary information on gene function and published information linked to PCa is provided here; more detailed information on each gene is provided in Table S1.

| Gene Name | Fraction | log FC | *p* Value | Expression | Gene Function/Link to Cancer |
|---|---|---|---|---|---|
| *CDKN3* | EVs | −2.9 | 0.0352 | Lower in PCa | Overexpressed in PCa [30] |
| *ERG_Exons_4-5* | EVs | 3.99 | 0.00650 | Higher in PCa | Overexpression due to TMPRSS2:ERG translocation [31] |
| *ERG_Exons_6-7* | Cell | 7.31 | $4.40 \times 10^{-6}$ | Higher in PCa | As above |
| *ERG_Exons_6-7* | EVs | 2.88 | 0.0342 | Higher in PCa | As above |
| *FOLH1* | Cell | 2.22 | 0.0474 | Higher in PCa | Overexpressed in PCa [32] |
| *HOXC6* | Cell | 3.45 | 0.0187 | Higher in PCa | Overexpressed in PCa urine sediment [33] |
| *HOXC6* | EVs | 1.65 | 0.0221 | Higher in PCa | Overexpressed in PCa urine sediment [33] |
| *OR51E2* | Cell | 3.27 | 0.0187 | Higher in PCa | Overexpressed in PCa urine sediment [34] |
| *PCA3* | EVs | 1.22 | 0.0306 | Higher in PCa | Overexpressed in PCa urine cell sediment [7] |
| *SLC12A1* | EVs | 2.99 | 0.0342 | Higher in PCa | Kidney-specific [35] |
| *SPINK1* | Cell | −3.12 | 0.0187 | Lower in PCa | Overexpressed in TMPRSS2:ERG-negative PCa [36] |
| *TMPRSS2:ERG* | EVs | 4.11 | 0.0893 | Higher in PCa | Translocation in 50% of PCa [31] |
| *UPK2* | Cell | −3.14 | 0.0187 | Lower in PCa | bladder-specific expression [37] |

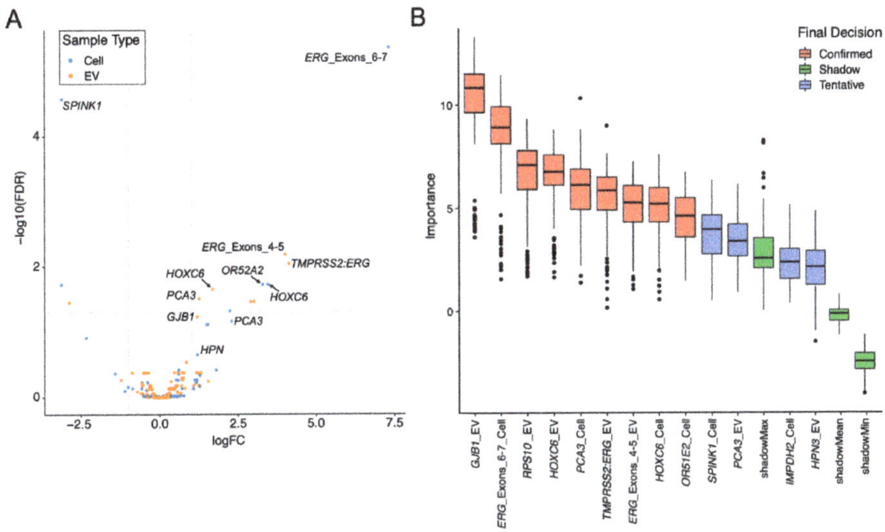


**Figure 2.** NanoString Gene-probe expression for cancer and non-cancer samples in urine cell-sediment (Cell) and extracellular vesicle (EV) samples. (**A**) Volcano plot, dashed lines are thresholds for significance (horizontal) and fold-changes (vertical). Genes selected by Boruta analysis are indicated. (**B**) Boruta selection of potentially useful gene-probes in prostate cancer detection. The fraction that the gene-probe was found to be useful in is indicated. Red indicates a gene-probe in the indicated urine fraction was significantly better than the ShadowMax feature ('confirmed'); blue indicates the gene data was significantly better than the ShadowMean ('tentative' see main text).

### *3.2. Expression Levels from RT-PCR and NanoString Are Strongly Correlated for Both EV and Cell Urine Fractions*

Quantitative RT-PCR (qRT-PCR) was used to verify the expression of five NanoString gene-probes: (i) *KLK3* a prostate-specific gene used for normalisation in the Progensa *PCA3* test [38]; (ii) *RPLP2*, a gene used for data normalisation in construction of the PUR signatures [10]; three commonly used PCa-related genes: (iii) *PCA3* (selected in Boruta analysis multivariate analysis for association with PCa—see below), (iv) *OR51E2* (aka *PSGR* Prostate-Specific G-Protein Coupled Receptor) (Boruta selected) and (v) *FOLH1* (aka PSMA, prostate-specific membrane antigen).

qRT-PCR Ct values for *RPLP2*, *FOLH1*, *OR51E2*, *PCA3* and *KLK3* were compared to the NanoString expression signals for these genes in 71 EV samples and 66 Cell samples (Figure S2). A strong correlation for *RPLP2*, *FOLH1*, *OR51E2* and *PCA3* was observed for both EV (Spearman correlation coefficient $r > 0.6$, $p < 0.00001$; Table 4) and Cell ($r > 0.6$, $p < 0.00001$; Table 4).

The *KLK3* data was more complex and a group of 13 samples (7xNC, 6xPCa) had low *KLK3* qRT-PCR/High Ct values in both EV and Cell fractions. When all the data were included, there was a strong correlation in Cell ($r = 0.70$, $p < 0.00001$) but in EV samples the correlation was weaker ($r = 0.51$, $p = 0.0017$) (See Discussion). Correlation of *KLK3* NanoString and RT-PCR data without these 13 samples provided $r$ values of >0.85 for both fractions.

Non-quantitative RT-PCR analysis was performed for the presence/absence of *TMPRSS2:ERG* fusion gene transcripts using *TMPRSS2* exon 1 forward and *ERG* exon 6 reverse primers. 14/21 PCa samples were positive for *TMPRSS2:ERG* by PCR in the EV fraction and 10/21 in the Cell fraction. The RT-PCR *TMPRSS2:ERG* status was significantly associated with the NanoString *TMPRSS2:ERG* values ($p = 4.36 \times 10^{-5}$ (EV); $p = 1.25 \times 10^{-4}$ (Cell); Mann-Whitney U test). In NC samples, RT-PCR also detected a *TMPRSS2:ERG* in

9/29 EV samples and 6/23 Cell samples. The data for NanoString probes *ERG_Exon_6-7* and *ERG_Exon_4-5* showed similar associations to RT-PCR positivity in both EV and Cell fractions (*ERG_Exon_6-7*: $p = 1.14 \times 10^{-5}$ (EV); $p = 1.51 \times 10^{-7}$ (Cell); *ERG_Exon_4-5*: $p = 3.40 \times 10^{-6}$ (EV); $p = 0.0113$ (Cell); Mann-Whitney U test). Nearly half (48%) of PCa EV samples were triple-positive for all three NanoString probes, while only 22% of Cell samples were triple-positive. Six of the 36 non-cancer samples (17%) were triple positive by EV but none were triple positive in Cell samples. All the triple-positive NC EV samples were from patients with a raised PSA and a PCa-negative-TRUS biopsy.

**Table 4.** Correlation results between qRT-PCR Ct values and NanoString expression signals.

| Gene | Experiment | Spearman Correlation Coefficient | *p*-Value |
|---|---|---|---|
| *FOLH1* | Cells | 0.71 | <0.00001 |
| *FOLH1* | EV | 0.68 | <0.00001 |
| *KLK3* | Cells | 0.7 | <0.00001 |
| *KLK3* | EV | 0.51 | <0.00001 |
| *OR51E2* | Cells | 0.77 | <0.00001 |
| *OR51E2* | EV | 0.74 | <0.00001 |
| *PCA3* | Cells | 0.88 | <0.00001 |
| *PCA3* | EV | 0.95 | <0.00001 |
| *RPLP2* | Cells | 0.86 | <0.00001 |
| *RPLP2* | EV | 0.79 | <0.00001 |

### *3.3. Each Urine Fraction Has Different Genes That Are Important in Predicting the Presence of Prostate Cancer*

The Boruta algorithm [16] was used to identify the importance of gene-probes in predicting the presence of PCa on biopsy. Thirteen gene-probe/urine fraction combinations were identified as performing significantly better than the mean performance of the Shadow gene-probe (ShadowMean) (see methods, Figure 2B). Nine of these gene-probe/urine fraction combinations were statistically better than the maximum performance of the Shadow feature (ShadowMax) and as such were deemed 'confirmed'. These nine gene-probe/urine fraction combinations corresponded to eight gene-probes providing readout from six genes (*GJB1, PCA3, HOXC6, OR51E2, RPS10, TMPRSS2:ERG*). Expression data for four examples are presented in Figure 3 (see Figure S1 for all Boruta-selected gene-probes).

#### 3.3.1. PCA3 and HOXC6 Were Useful in Both EV and Cell Sediment Fractions

Two gene-probes were identified as being useful in both Cell and EV fractions: *HOXC6* and *PCA3*.

*PCA3* (Prostate Cancer Associated 3, Figure 3C) was a potentially useful feature in both fractions, albeit tentatively in the EV fraction. In the Cells, 36% of the NC samples had no expression of *PCA3* compared to 8% of the PCa samples. The median expression of *PCA3* in PCa v NC samples was 8.6-fold higher in the Cell fraction and 2.6-fold higher in the EV. *PCA3* is a prostate-tissue-specific, noncoding messenger RNA [39] overexpressed in urine cell sediment in 95% of men with PCa [7]. A *PCA3* assay has been developed using *PCA3* expression in urine cell-sediment (Progensa®; Gen-Probe, San Diego, CA, USA). The *PCA3* test has been approved by the FDA as a diagnostic test only in the setting of an initial negative prostate biopsy to predict the presence of PCa on a second biopsy [7]. The PCA3 test has not been approved for use in the National Health Service in the UK [2] and the European Association of Urology has stated that its impact at a single-patient level remains highly questionable [40]. In tissue, *PCA3* has a bimodal distribution in both biopsy and radical prostatectomy (RP) samples, where low *PCA3* expression was significantly associated with high grade disease ($p < 0.001$). *PCA3* had a poor performance in predicting high grade disease in initial biopsy tissue (GS $\geq$ 8) with 55% sensitivity and high false negative rates. In excised prostates, low *PCA3* is also associated with adverse pathological features, clinical recurrence outcome and a greater probability of metastatic progression ($p < 0.001$) [41]. In meta-analyses of PCA3-test studies of patients with previous negative biopsies, an AUC of 0.739 (with a *PCA3* score cut-off of >35) and an AUC of 0.63 were found for PCa on a second biopsy [42]. Meta-analyses of the urine test by Luo et al. (2014) concluded that results were heterogeneous (sensitivity: 0.75–0.93; specificity 0.44–0.78) [43].

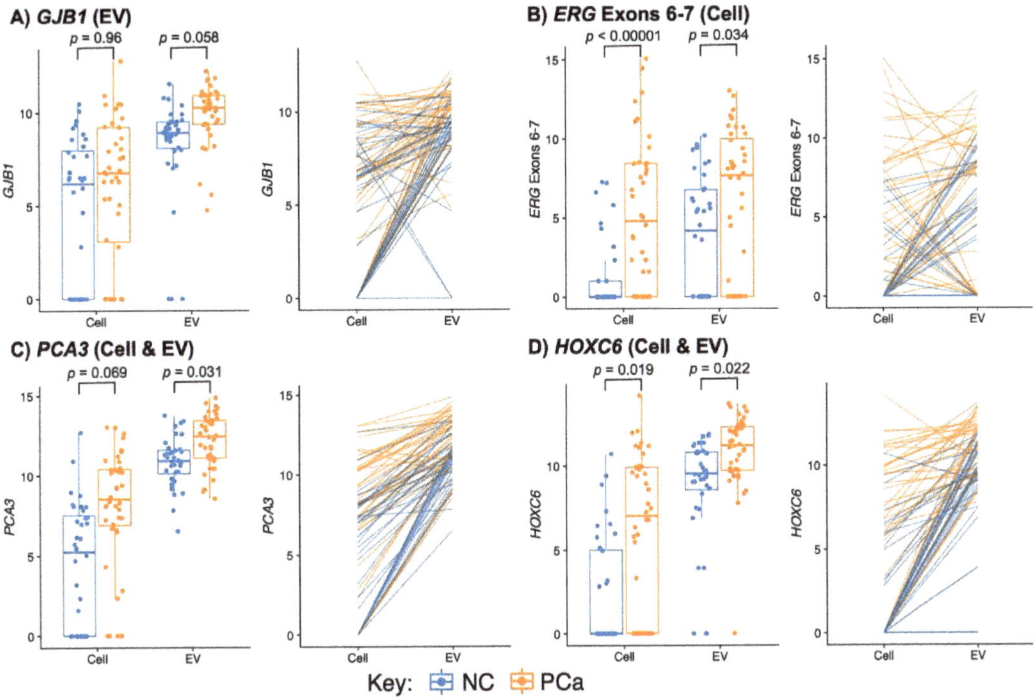


**Figure 3.** Box plots of Boruta-selected NanoString gene-probe data from urine cell sediment (Cell) and urine extracellular vesicle (EV) RNA in men with prostate cancer (PCa, orange) and controls with no evidence of cancer (NC, blue, see Methods). Orange (PCa) and blue (NC) lines link NanoString expression data for paired Cell and EV samples from individual urine samples. *p*-values (False Discovery Rate) are for the statistical difference between cancer and non-cancer by edgeR. Gene-probes were: (**A**) *GJB1*, (**B**) *ERG*_Exons_6-7, (**C**) *PCA3*, (**D**) *HOXC6*.

*HOXC6* (Homeobox C6, Figure 3D) had much higher expression levels in EV than Cell (median 1432 vs. median 8). In EV, *HOXC6* was expressed by >95% of the samples with median expression 3·2-fold higher in the PCa samples than in NC. In the Cell fraction, *HOXC6* was detected in 68% of the PCa samples compared to 36% of the NC samples. *HOXC6* is overexpressed in primary, metastasized and castration-resistant PCa, and expression was not influenced by androgens or treatments targeting the AR signaling pathway [33]. Silencing of *HOXC6* expression using small-interfering RNA (siRNA) resulted in decreased proliferation rates for both androgen-dependent LnCaP cells and the LnCaP- derived androgen-independent C4-2 cell line, and induced apoptosis [44]. *HOXC6* mRNA levels are higher in the urine cell-sediment of PCa patients [33], and patients with high *HOXC6* expression had shorter overall survival than those with low *HOXC6* expression [45]. *HOXC6* is used in the SelectMDx prostate cancer urine test alongside *DLX1*. SelectMDx has been found to underperform when compared to template biopsy [46] and mpMRI [47] in the detection of clinically significant PCa.

3.3.2. EV Fraction Genes Useful for Prostate Cancer Detection

Five gene-probes were useful in the EV fraction only: *GJB1*, *RPS10*, *TMPRSS2:ERG*, *ERG*_Exons_4-5, and *HPN*. Data for *TMPRSS2:ERG* and the two *ERG* probes are presented in Section 3.3.4.

*GJB1* (Gap Junction Protein Beta 1) expression in EVs (Figure 3A) was identified with the highest importance in discerning PCa from NC. *GJB1* expression in EVs was a median of 2.5-fold lower in the NC samples compared to the PCa samples and 70% (25/36) of the NC samples had expression below the lowest quartile of the PCa samples. In contrast, expression of *GJB1* in the Cell was not significantly different in cancer and non-cancer samples. *GJB1* has been associated with PCa [48] and has been identified as a prognostic marker for renal cancer [49]. *GJB1* is a member of the gap junction connexin family of proteins that regulates and controls the transfer of communication signals across cell membranes, primarily in the liver and peripheral nervous system. Expression levels of GJB1 protein (aka Connexin 32, CX32) were found to be the same in PCa and benign prostatic hyperplasia samples [50]. No publications for the use of *GJB1* in urine tests were found apart from one publication by our group (Connell et al., 2020 [17]).

*RPS10* (Ribosomal Protein S10) was highly expressed in both fractions and no samples were negative. It was identified as being useful in detecting PCa within the EV fraction with *RPS10* expression levels decreased in cancer (Figure S1). *RPS10* has been found to be overexpressed at the protein level in PCa [51]. We have not found any reports suggesting the use of *RPS10* as a urinary biomarker.

*HPN* (Hepsin) was tentatively identified as being potentially useful for PCa-detection and therefore would require further testing in a larger cohort. It encodes a type II transmembrane serine protease involved in diverse cellular functions, including the maintenance of cell morphology. HPN is upregulated in PCa and correlates with disease progression [52].

3.3.3. Cellular Genes Useful for PCa Detection

Four gene-probes were useful in the Cell fraction only: *ERG*_Exons_6-7, *OR51E2*, *SPINK1* and *IMPDH2*.

*OR51E2* (Olfactory Receptor Family 51 Subfamily E Member 2, Figure S1) was 30-fold more highly expressed in the EV fraction compared to the Cell fraction (median 2006 vs. 63). However, *OR51E2* was only useful for detecting PCa in the Cell fraction, in which NanoString signal was above the threshold in only 33% of NC compared to 73% of PCa, with a median differential expression of 127-fold. *OR51E2* has been found to be overexpressed in PCa tissue [53] and in the urine cell-sediment of men with PCa [34].

*IMPDH2* followed a similar pattern of expression to *OR51E2*, being higher in EVs but more informative in the Cell fraction for PCa detection. *IMPDH2* (Inosine Monophosphate Dehydrogenase 2) encodes the rate-limiting enzyme in the de novo guanine nucleotide biosynthesis required for DNA and RNA synthesis. Increased serum levels of *IMPDH2* were significantly associated with Gleason $\geq$8 PCa, suggesting its potential as a serological tumor marker [54]. *IMPDH2*, in our study, was identified as being potentially useful but only tentatively and would require further testing in a larger cohort.

*SPINK1* (Serine Peptidase Inhibitor Kazal Type 1) was the only Boruta-selected probe that had a higher median expression in Cell than EVs (~2-fold). *SPINK1* has been reported to be overexpressed in a group of *ETS*-fusion negative PCa and *SPINK1*-positive PCa was reported to be an aggressive PCa subtype [36]. Laxman et al. [55] demonstrated an increase in *SPINK1* in PCa and suggested its use in a multiplex assay using urinary sediments.

3.3.4. TMPRSS2:ERG and ERG Probes Are Useful Biomarkers in Urine

A *TMPRSS2:ERG* translocation is detectable in ~50% of all prostate cancer foci [33,56], and results in overexpression of *ERG* (ETS Transcription Factor ERG) from exon 4 to 3′-end in >95% of TMPRSS2:ERG cases [57]. An increased copy number of *TMPRSS2:ERG* has been associated with a worse prognosis [56]. Three NanoString probes were designed to detect transcripts from a *TMPRSS2:ERG* fusion gene: a *TMPRSS2:ERG* gene probe spanning the most commonly found *TMPRSS2*_exon1/*ERG*_exon4 fusion transcripts [57] and two probes to *ERG* sequences that lay 3′ to the usual gene translocation point, one spanning exons 4 to 5 (*ERG*_Exon4-5), and the other spanning exons 6 to 7 (*ERG*_Exon_6-7). All three *ERG* probes were found to be useful for PCa detection: *ERG*_Exons_6-7 levels in Cell (Figure 3) and EV levels of *ERG*_Exons_4-5 and *TMPRSS2:ERG* fusion (Figure S1). EV expression levels for the *TMPRSS2:ERG* and *ERG*_Exons_4-5 probes in PCa samples were similar to each other and were ~4.5 times higher than probe signals in the Cells. Median signals for the *ERG*_Exon_6-7 probe were much higher in both fractions than those obtained from the *TMPRSS2:ERG* and *ERG*_Exon_4-5 probes: ~2.5-fold higher in the EV, and 6.5-fold higher in the Cell fractions (see Figure S2 and Discussion). In the PCa EV samples, the three probes—'*TMPRSS2:ERG*', '*ERG*_Exon_4-5' and '*ERG*_Exon_6-7' probe—were above the threshold in 22/40, 23/40 and 26/40 PCa samples, respectively and 95% of the *TMPRSS2:ERG*-positive samples were positive for *ERG*_Exon_4-5 and 90% for *ERG*_Exon_6-7, with 19/40 PCa samples triple-positive for all three probes in the EV fraction. For the Cell PCa samples, the three probes had lower rates of detection (11/40, 10/40, 23/40) with 81% *ERG*_Exon_4-5 and 100% *ERG*_Exon_6-7 in concurrence with the *TMPRSS2:ERG* probe positive samples. In addition, 9/40 PCa samples were triple-positive in the Cell fraction. For the NC samples, 6/36 were triple positive in the EV fraction, but none were triple positive in the Cell fraction. All the triple-positive NC samples had a raised PSA (>4 ng/mL) but were negative for PCa on biopsy.

Due to the multifocal nature of PCa, tumor foci can be found with and without a *TMPRSS2:ERG* in individual prostates [58] such that they are present in ~70% of cancerous prostates [59], making them a more useful biomarker than was originally apparent. Young et al. determined that *TMPRSS2:ERG* urine transcript levels aided *PCA3* in predicting the presence of PCa and correlated with *ERG* expression in PCa tissue [60]. Tomlins et al. combined the detection of *TMPRSS2:ERG* fusion transcripts and *PCA3* with serum PSA levels and the result from the multivariate Prostate Cancer Prevention Trial risk calculator version 1.0 (PCPT-RC) in a combined predictor, which they called Mi-Prostate score (MiPS) [61]. MiPS had a significantly improved AUC for the detection of PCa and Gs $\geq 7$ on biopsy when compared to PSA or PCPT-RC alone.

### 3.4. Gene-Transcript Expression in Urine EV and Urine Cell-Sediment Are Equally Useful for Detecting Prostate Cancer

A random forests model to predict cancer status was built incorporating the gene-probes identified by Boruta analysis for the samples in each fraction (Section 3.2); in addition, an optimal predictor was produced for the EV and Cell fractions combined (Table S4). The output for each model was a diagnostic risk score. Each of the three signatures were able to predict the presence of cancer, with the area under the curve values (AUCs) being significantly better than a random predictor. AUCs for the three models were: EV signature AUC 0.82 (bootstrap Confidence Interval 0.729–0.921), Cell signature 0.79 (0.684–0.894), Combined signature (0.87 (0.788–0.944), Figure 4D). The combined model had the highest AUC, which was within the range of the confidence intervals for the other signatures and so there was no evidence for a significant improvement. Density plots for the three models were constructed (Figure 4A–C), each signature showing distinct peaks for NC and PCa.

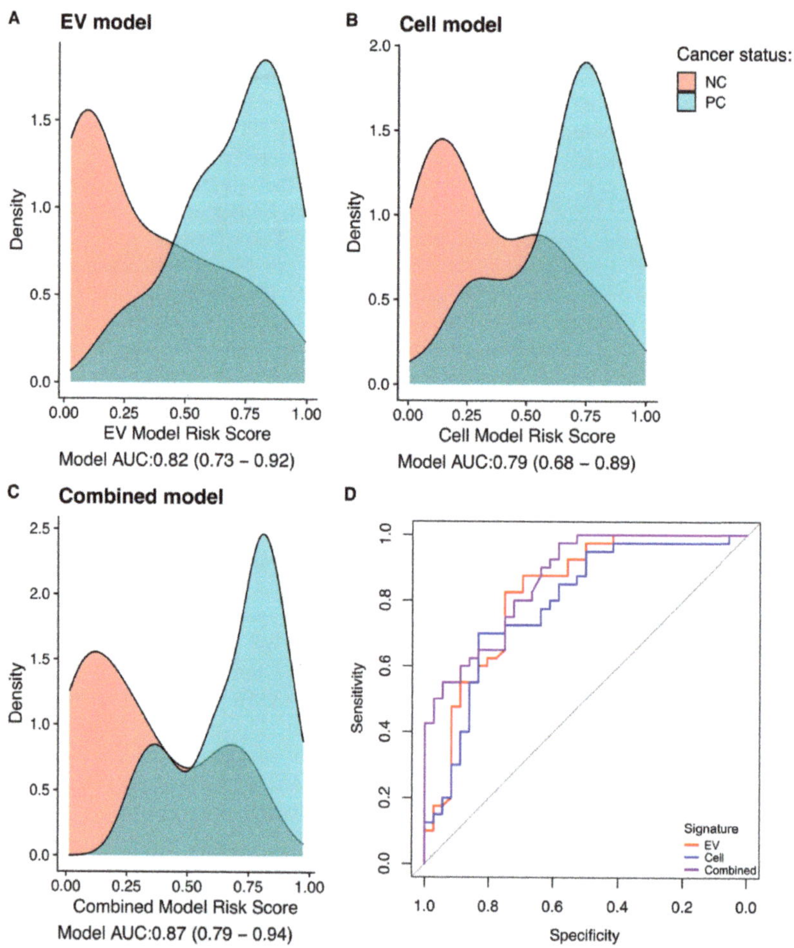


**Figure 4.** Model Metrics. (**A–C**): Density plots for the EC, Cell, and combined signatures (models). Red is non-cancer and blue is prostate cancer. (**D**): AUC analysis for the three signatures: red = EV, blue = Cell, purple = combined signature.

## 4. Discussion

We have examined the transcriptomes of urine EVs and cell-sediment from 76 men with a large number of gene-probes ($n = 167$). We have compared the relative expression of these gene-probes in 100 ng of amplified cDNA products from each fraction and investigated their usefulness in Pca detection. Thirteen gene-probe/urine fraction combinations were identified as being potentially useful in the identification of prostate cancer. *GJB1* expression in EVs was found to be the strongest candidate. Five gene-probes were solely useful in the EV fraction and four gene-probes were solely useful in the cell-sediment, suggesting that fractionation prior to analysis can provide more potential biomarkers. Only *PCA3* and *HOXC6* were useful in both fractions. Three models were constructed from the EV, Cell and combined EV & Cell data which showed a strong separation of PCa and non-cancer samples.

The vast majority of the NanoString probes used here were designed for gene transcripts reported to be differentially expressed in PCa tissue [10]. It was therefore surprising that the bulk of the 154 PCa-linked gene probes did not show any useful association with cancer in urine cell-sediment or EV RNA. The potential reasons for this are different for each fraction. In cell sediment, PCa cells are a tiny minority [62], data which is supported by the very strong expression of the nucleated-blood-cell gene PTPRC in urine cells by Quek [63] and by our data here in both PCa and non-cancer urine samples. We and Pellegini et al. [12] observed a much higher level of expression of prostate-specific genes (*KLK2* and *KLK3*) in the EV fraction than the Cell fraction (*KLK2* 22-fold higher, *KLK3* 50-fold higher) indicating an enrichment for prostate-specific transcripts within the EV fraction compared to the cell sediments (Figure 1).

Boruta analysis of the two urinary fractions identified 13 gene-probe/urine fraction combinations as being potentially useful in identifying prostate cancer. These selected genes included *PCA3*, *HOXC6* and *TMPRSS2:ERG*, that have previously been identified as having utility as urinary biomarkers for PCa [61,64–66]. Two of these genes (*PCA3* and *HOXC6*) appear to be useful in both the Cell and EV fractions, though the Boruta importance score was higher for the EV fraction. Expression was observed in almost all EV samples (Figure 3) in comparison to for example *HOXC6* which had limited expression in the NC samples being detected in only 14 of 35 samples (39%) compared to 28 of 40 (70%) in the PCa groups. The gene identified as having the most predictive utility was *GJB1* in the EV fraction (Figure 2). *GJB1* was highly expressed by almost all the samples in the EV fraction with a higher median gene expression in the PCa-associated groups. Three NanoString probes were designed to detect *TMPRSS2:ERG* fusion transcripts or overexpression of *ERG* as a result of the translocation [19,31]. The robustness of detecting *TMPRSS2:ERG* in samples was very different between the EV and Cell fractions and there was a lack of sensitivity for the detection of *TMPRSS2:ERG* in cell-sediment, which may reflect the low levels of PCa cells in this fraction. Patients with a PCa-negative 10-core TRUS biopsy have been reported to harbour undiscovered PCa in around 20% of cases [67]. The *ERG*_Exons_6-7 probe appeared to have a much higher sensitivity of detection than the other two probes, with *ERG*_Exons_6-7 detecting raised expression of *ERG* in 23 Cell samples compared to *ERG*_Exons_4-5 ($n$ = 10) or *TMPRSS2:ERG* ($n$ = 11). The reason for this may relate to the precise fusion transcripts generated or the extremely GC-rich nature of *TMPRSS2* exon1 sequences (79% GC). GC-rich regions have a much poorer efficiency of reverse transcription [68] due to RNA secondary structure which could result in a lower detection rate for the *TMPRSS2* sequences and the immediately adjacent *ERG*_Exons_4-5 sequence relative to the more distal *ERG* ex 6-7 sequences. The *ERG*_Exons_6-7 probe gave on average a 2.5-fold higher signal than the other two probes in EVs and a 6.5-fold higher signal in the Cell fractions of PCa samples. We hypothesise that it was this additional sensitivity of the *ERG*_Exons_6-7 probe that enabled it to have utility in the Cell fraction.

Comparison of qRT-PCR data with NanoString data for four genes displayed a strong correlation between the two methods indicating that NanoString is a useful method for multiplex gene expression analysis in agreement with previous studies [69]. In contrast, the correlation between expression values from NanoString and qRT-PCR for *KLK3* was poor, which was due to low expression values for *KLK3* by qRT-PCR in a subgroup of samples, a difference that was not detected by NanoString. Interestingly, the *KLK3* qRT-PCR values were low in both Cell and EV fractions of the same 13 urine samples. A possible explanation for this comes from David et al. 2002, who found that transcript splice variants of *KLK3* can include all or part of intron 1 [70]. Notably, the forward *KLK3* PCR primer used in our investigations spanned the exon1-2 boundary, with the two 3′ nucleotides being in exon 2. Thus, the presence of intron 1 sequences would make the transcripts un-PCRable with this primer set. The NanoString probes are much larger (2 × 50 nt) and reported *KLK3* levels in these samples were not discernible from the other samples.

In our study presented here, three gene probes (*GJB1*, *RPS10* and *HOXC6*) provided comparable utility to *PCA3*. *GJB1* and *RPS10* have not to our knowledge been used in urine tests by other laboratories and could open up new avenues of research. Our data suggest that increasing the number of PCa-associated genes in a urine test should provide a more level playing field for the detection of cancers. Knowing which urine fraction to use for these multi-gene tests is critical. In addition, our results demonstrate that the exact probe sequences used to detect gene transcripts expressed by, for example, a *TMPRSS2:ERG* fusion gene is important.

We identified different genes as useful biomarkers in different urinary fractions indicating the utility of using both fractions for biomarker development. Using whole urine is attractive due to the reduction in required preparation steps and Hendriks et al. [13] suggested that whole urine is a useful substrate, at least for *PCA3* and *ERG*. In our results, *PCA3* and *ERG* probes were highlighted as potential biomarkers in both the Cell and EV fractions. However, using whole urine for other novel markers may reduce the potential ability to detect PCa. For example, *OR51E2* is ~10-fold more highly expressed in EVs than the Cell fraction (Figure S1) but was only useful as a PCa-biomarker in the Cell fraction, therefore if the two fractions were combined it is likely that the high EV expression would obscure the difference in expression between the NC and PCa groups seen in the cell-sediments. Similarly, for *RPS10* the reduction in expression in the PCa groups of the EV fraction may be lost when combined with the expression from the Cell fraction. These data would suggest that screens for new PCa-biomarkers should be conducted on each individual fraction and that multiple probes for the detection of individual gene transcripts should be tested to optimise performance. An aspect not covered in this study is that of gene mutations, for example, mutations of mitochondrial genes associated with patient survival [71], and it may be fruitful to integrate targeted analysis of specific urine gene transcripts for mutations in future urine tests. We are currently in the validation phase of our urine research for which we have collected 2500 samples for analysis and are creating an accredited diagnostics laboratory that will enable us to provide urine results to patients.

## 5. Conclusions

We have interrogated urine Cell and EV RNA with 167 gene-probes and observed differential expression between fractions. We have identified 11 genes as useful in identifying PCa, which are distributed between the Cell and EV fractions, including the biomarker *GJB1*. These data indicate that a useful strategy for improving the identification of PCa through urinary biomarker analysis would involve the measurement of specific gene-targets from different urinary fractions.

**Supplementary Materials:** The following supporting information can be downloaded at: https://www.mdpi.com/article/10.3390/cancers15030789/s1, Figure S1: NanoString expression data for Boruta selected genes; Figure S2: Comparison of NanoString data and qRT-PCR data for 5 gene-transcripts; Table S1: Nanostring Data; Table S2: Differential gene-expression analysis; Table S3: RT-PCR-Nanostring comparisons; Table S4: Boruta signatures.

**Author Contributions:** Conceptualization, J.C., C.S.C., M.Y.H., S.P.O.; Methodology, M.Y.H., J.C, M.O., R.H.; Software, M.W., S.P.O., D.S.B.; Formal Analysis, M.W., S.P.O., D.S.B. J.C.; Investigation, M.Y.H., M.O., R.H.; Resources, R.M., R.H., M.Y.H.; Data Curation, D.S.B.; Writing—Original Draft Preparation, M.W., J.C.; Writing—Review & Editing, J.C., D.S.B., C.S.C., R.M., R.H., M.O., M.W., S.P.O., M.Y.H.; Visualization, J.C., M.W., D.S.B.; Supervision, J.C., C.S.C., D.S.B.; Project Administration, J.C.; Funding Acquisition, C.S.C., D.S.B. All authors have read and agreed to the published version of the manuscript.

**Funding:** This study was possible thanks to the Movember Foundation (GAP1) Urine Biomarker project, a Prostate Cancer UK and Movember Foundation Research Innovation Award (RIA18-ST2-014), The Masonic Charitable Foundation, The Bob Champion Cancer Trust, the King family, Big C Cancer Charity, The Andy Ripley Memorial Fund and the Stephen Hargrave Trust. The sponsors had no role in the design, execution, interpretation, or writing of the study.

**Institutional Review Board Statement:** The study was conducted according to the guidelines of the Declaration of Helsinki, and approved by the Health Research Authority and the NRES Committee East of England Ethics Committee, 2012. REC reference: 12/EE/0058, IRAS project ID: 196199.

**Informed Consent Statement:** Informed consent was obtained from all subjects involved in the study.

**Data Availability Statement:** Data is contained within the article or supplementary material.

**Acknowledgments:** The authors would like to thank those men with prostate cancer and the subjects who have donated their time and samples to the data sets used in this study. Part of the research presented in this paper was carried out on the High-Performance Computing Cluster supported by the Research and Specialist Computing Support service at the University of East Anglia.

**Conflicts of Interest:** A patent application has been filed by C.S.C., J.C. and D.S.B. for the use of the PUR biomarkers in prostate cancer diagnosis and prognosis. There are no other conflict of interest to disclose.

## References

1. Bray, F.; Ferlay, J.; Soerjomataram, I.; Siegel, R.L.; Torre, L.A.; Jemal, A. Global cancer statistics 2018: GLOBOCAN estimates of incidence and mortality worldwide for 36 cancers in 185 countries. *CA Cancer J. Clin.* **2018**, *68*, 394–424. [CrossRef]
2. National Institute for Health and Care Excellence (NICE). *Prostate Cancer: Diagnosis and Management*; National Institute for Health and Care Excellence (NICE): London, UK, 2021; pp. 1–58.
3. Bernemann, C.; Schnoeller, T.J.; Luedeke, M.; Steinestel, K.; Boegemann, M.; Schrader, A.J.; Steinestel, J. Expression of AR-V7 in Circulating Tumour Cells Does Not Preclude Response to Next Generation Androgen Deprivation Therapy in Patients with Castration Resistant Prostate Cancer. *Eur. Urol.* **2017**, *71*, 1–3. [CrossRef]
4. Hennigan, S.T.; Trostel, S.Y.; Terrigino, N.T.; Voznesensky, O.S.; Schaefer, R.J.; Whitlock, N.C.; Wilkinson, S.; Carrabba, N.V.; Atway, R.; Shema, S.; et al. Low Abundance of Circulating Tumor DNA in Localized Prostate Cancer. *JCO Precis. Oncol.* **2019**, *3*, 1–13. [CrossRef] [PubMed]
5. Herbut, P.A.; Lubin, E. Cancer Cells in Prostatic Secretions. *J. Urol.* **1947**, *57*, 542–551. [CrossRef] [PubMed]
6. Mitchell, P.J.; Welton, J.; Staffurth, J.; Court, J.; Mason, M.D.; Tabi, Z.; Clayton, A. Can urinary exosomes act as treatment response markers in prostate cancer? *J. Transl. Med.* **2009**, *7*, 4. [CrossRef] [PubMed]
7. Hessels, D.; Gunnewiek, J.M.K.; van Oort, I.; Karthaus, H.F.; van Leenders, G.J.; van Balken, B.; Kiemeney, L.; Witjes, J.; Schalken, J.A. DD3PCA3-based Molecular Urine Analysis for the Diagnosis of Prostate Cancer. *Eur. Urol.* **2003**, *44*, 8–16. [CrossRef]
8. Hessels, D.; Smit, F.P.; Verhaegh, G.W.; Witjes, J.A.; Cornel, E.B.; Schalken, J.A. Detection of TMPRSS2-ERG Fusion Transcripts and Prostate Cancer Antigen 3 in Urinary Sediments May Improve Diagnosis of Prostate Cancer. *Clin. Cancer Res.* **2007**, *13*, 5103–5108. [CrossRef]
9. Miranda, K.C.; Bond, D.T.; McKee, M.; Skog, J.; Păunescu, T.G.; Da Silva, N.; Brown, D.; Russo, L.M. Nucleic acids within urinary exosomes/microvesicles are potential biomarkers for renal disease. *Kidney Int.* **2010**, *78*, 191–199. [CrossRef]
10. Connell, S.P.; Yazbek-Hanna, M.; McCarthy, F.; Hurst, R.; Webb, M.; Curley, H.; Walker, H.; Mills, R.; Ball, R.Y.; Sanda, M.G.; et al. A four-group urine risk classifier for predicting outcomes in patients with prostate cancer. *BJU Int.* **2019**, *124*, 609–620. [CrossRef]
11. Dijkstra, S.; Birker, I.L.; Smit, F.P.; Leyten, G.H.J.M.; De Reijke, T.M.; Van Oort, I.M.; Mulders, P.F.A.; Jannink, S.A.; Schalken, J.A. Prostate Cancer Biomarker Profiles in Urinary Sediments and Exosomes. *J. Urol.* **2014**, *191*, 1132–1138. [CrossRef]
12. Pellegrini, K.L.; Patil, D.; Douglas, K.J.; Lee, G.; Wehrmeyer, K.; Torlak, M.; Clark, J.; Cooper, C.S.; Moreno, C.S.; Sanda, M.G. Detection of prostate cancer-specific transcripts in extracellular vesicles isolated from post-DRE urine. *Prostate* **2017**, *77*, 990–999. [CrossRef] [PubMed]
13. Hendriks, R.J.; Dijkstra, S.; Jannink, S.A.; Steffens, M.G.; van Oort, I.M.; Mulders, P.F.; Schalken, J.A. Comparative analysis of prostate cancer specific biomarkers PCA3 and ERG in whole urine, urinary sediments and exosomes. *Clin. Chem. Lab. Med.* **2016**, *54*, 483–492. [CrossRef] [PubMed]
14. DeAntoni, E.P.; Crawford, E.D.; Oesterling, J.E.; Ross, C.A.; Berger, E.R.; McLeod, D.G.; Staggers, F.; Stone, N.N. Age- and race-specific reference ranges for prostate-specific antigen from a large community-based study. *Urology* **1996**, *48*, 234–239. [CrossRef]
15. Chen, Y.; Lun, A.T.L.; Smyth, G.K. From reads to genes to pathways: Differential expression analysis of RNA-Seq experiments using Rsubread and the edgeR quasi-likelihood pipeline. *F1000Research* **2016**, *5*, 1438. [CrossRef]
16. Kursa, M.B.; Rudnicki, W.R. Feature Selection with the Boruta Package. *J. Stat. Softw.* **2010**, *36*, 1–13. [CrossRef]
17. Connell, S.P.; O'Reilly, E.; Tuzova, A.; Webb, M.; Hurst, R.; Mills, R.; Zhao, F.; Bapat, B.; Cooper, C.S.; Perry, A.S.; et al. Development of a multivariable risk model integrating urinary cell DNA methylation and cell-free RNA data for the detection of significant prostate cancer. *Prostate* **2020**, *80*, 547–558. [CrossRef]
18. Sequeiros, T.; Bastarós, J.M.; Sánchez, M.; Rigau, M.; Montes, M.; Placer, J.; Planas, J.; de Torres, I.; Reventós, J.; Pegtel, D.M.; et al. Urinary biomarkers for the detection of prostate cancer in patients with high-grade prostatic intraepithelial neoplasia. *Prostate* **2015**, *75*, 1102–1113. [CrossRef]
19. Clark, J.; Merson, S.; Jhavar, S.; Flohr, P.; Edwards, S.; Foster, C.S.; Eeles, R.; Martin, F.L.; Phillips, D.H.; Crundwell, M.; et al. Diversity of TMPRSS2-ERG fusion transcripts in the human prostate. *Oncogene* **2006**, *26*, 2667–2673. [CrossRef]
20. Cohen, J. *Statistical Power Analysis for the Behavioral Sciences*; Routledge: New York, NY, USA, 2013; ISBN 0-8058-0283-5. [CrossRef]
21. Wen, S.; Wei, Y.; Zen, C.; Xiong, W.; Niu, Y.; Zhao, Y. Long non-coding RNA NEAT1 promotes bone metastasis of prostate cancer through N6-methyladenosine. *Mol. Cancer* **2020**, *19*, 171. [CrossRef]

22. Gervais, F.G.; Singaraja, R.; Xanthoudakis, S.; Gutekunst, C.-A.; Leavitt, B.R.; Metzler, M.; Hackam, A.S.; Tam, J.; Vaillancourt, J.P.; Houtzager, V.; et al. Recruitment and activation of caspase-8 by the Huntingtin-interacting protein Hip-1 and a novel partner Hippi. *Nature* **2002**, *4*, 95–105. [CrossRef]
23. De Jonge, H.J.M.; Fehrmann, R.; De Bont, E.S.J.M.; Hofstra, R.; Gerbens, F.; Kamps, W.A.; de Vries, E.; Van Der Zee, A.G.J.; Meerman, G.J.T.; Ter Elst, A. Evidence Based Selection of Housekeeping Genes. *PLoS ONE* **2007**, *2*, e898. [CrossRef] [PubMed]
24. Mao, B.; Zhang, Z.; Wang, G. BTG2: A rising star of tumor suppressors (Review). *Int. J. Oncol.* **2014**, *46*, 459–464. [CrossRef]
25. Shabani, S.; Elahi, E.; Bahraniasl, M.; Babaheidarian, P.; Sadeghpour, A.; Majidzadeh, T.; Talebi, A.; Mahjoubi, F. Multi-stage analysis of FOXM1, PYROXD1, hTERT, PPARA, PIM3, BMI1 and MCTP1 expression patterns in colorectal cancer. *Carcinogenesis* **2022**, *9*, 10.
26. Lu, Z.; Qi, L.; Bo, X.J.; Liu, G.D.; Wang, J.M.; Li, G. Expression of CD26 and CXCR4 in prostate carcinoma and its relationship with clinical parameters. *J. Res. Med. Sci.* **2013**, *18*, 647–652. [PubMed]
27. Su, W.-P.; Sun, L.-N.; Yang, S.-L.; Zhao, H.; Zeng, T.-Y.; Wu, W.-Z.; Wang, D. Apolipoprotein C1 promotes prostate cancer cell proliferation in vitro. *J. Biochem. Mol. Toxicol.* **2018**, *32*, e22158. [CrossRef] [PubMed]
28. Yan, N.; Shi, Y. Histone H1.2 as a trigger for apoptosis. *Nat. Struct. Mol. Biol.* **2003**, *10*, 983–985. [CrossRef] [PubMed]
29. Itkonen, H.M.; Brown, M.; Urbanucci, A.; Tredwell, G.; Lau, C.H.; Barfeld, S.; Hart, C.; Guldvik, I.J.; Takhar, M.; Heemers, H.V.; et al. Lipid degradation promotes prostate cancer cell survival. *Oncotarget* **2017**, *8*, 38264–38275. [CrossRef]
30. Yu, C.; Cao, H.; He, X.; Sun, P.; Feng, Y.; Chen, L.; Gong, H. Cyclin-dependent kinase inhibitor 3 (CDKN3) plays a critical role in prostate cancer via regulating cell cycle and DNA replication signaling. *Biomed. Pharmacother.* **2017**, *96*, 1109–1118. [CrossRef]
31. Tomlins, S.A.; Rhodes, D.R.; Perner, S.; Dhanasekaran, S.M.; Mehra, R.; Sun, X.-W.; Varambally, S.; Cao, X.; Tchinda, J.; Kuefer, R.; et al. Recurrent Fusion of *TMPRSS2* and ETS Transcription Factor Genes in Prostate Cancer. *Science* **2005**, *310*, 644–648. [CrossRef]
32. Maraj, B.; Markham, A. Prostate-specific membrane antigen (FOLH1): Recent advances in characterising this putative prostate cancer gene. *Prostate Cancer Prostatic Dis.* **1999**, *2*, 180–185. [CrossRef]
33. Hamid, A.R.A.H.; Hoogland, A.M.; Smit, F.; Jannink, S.; van Rijt-van de Westerlo, C.; Jansen, C.F.J.; van Leenders, G.J.L.H.; Verhaegh, G.W.; Schalken, J.A. The role of HOXC6 in prostate cancer development. *Prostate* **2015**, *75*, 1868–1876. [CrossRef] [PubMed]
34. Rigau, M.; Morote, J.; Mir, M.C.; Ballesteros, C.; Ortega, I.; Sanchez, A.; Colás, E.; Garcia, M.; Ruiz, A.; Abal, M.; et al. PSGR and PCA3 as biomarkers for the detection of prostate cancer in urine. *Prostate* **2010**, *70*, 1760–1767. [CrossRef] [PubMed]
35. Haas, M. The Na-K-Cl cotransporters. *Am. J. Physiol. Physiol.* **1994**, *267*, C869–C885. [CrossRef]
36. Tomlins, S.A.; Rhodes, D.R.; Yu, J.; Varambally, S.; Mehra, R.; Perner, S.; Demichelis, F.; Helgeson, B.E.; Laxman, B.; Morris, D.S.; et al. The Role of SPINK1 in ETS Rearrangement-Negative Prostate Cancers. *Cancer Cell* **2008**, *13*, 519–528. [CrossRef] [PubMed]
37. Lobban, E.D.; Smith, B.A.; Hall, G.D.; Harnden, P.; Roberts, P.; Selby, P.J.; Trejdosiewicz, L.K.; Southgate, J. Uroplakin Gene Expression by Normal and Neoplastic Human Urothelium. *Am. J. Pathol.* **1998**, *153*, 1957–1967. [CrossRef] [PubMed]
38. Groskopf, J.; Aubin, S.M.; Deras, I.L.; Blase, A.; Bodrug, S.; Clark, C.; Brentano, S.; Mathis, J.; Pham, J.; Meyer, T.; et al. APTIMA PCA3 Molecular Urine Test: Development of a Method to Aid in the Diagnosis of Prostate Cancer. *Clin. Chem.* **2006**, *52*, 1089–1095. [CrossRef]
39. De Kok, J.B.; Verhaegh, G.W.; Roelofs, R.W.; Hessels, D.; Kiemeney, L.A.; Aalders, T.W.; Swinkels, D.W.; Schalken, J.A. DD3(PCA3), a very sensitive and specific marker to detect prostate tumors. *Cancer Res.* **2002**, *62*, 2695–2698. [PubMed]
40. Heidenreich, A.; Bellmunt, J.; Bolla, M.; Joniau, S.; Mason, M.; Matveev, V.; Mottet, N.; Schmid, H.-P.; van der Kwast, T.; Wiegel, T.; et al. EAU Guidelines on Prostate Cancer. Part 1: Screening, Diagnosis, and Treatment of Clinically Localised Disease. *Eur. Urol.* **2011**, *59*, 61–71. [CrossRef]
41. Alshalalfa, M.; Verhaegh, G.W.; Gibb, E.A.; Santiago-Jiménez, M.; Erho, N.; Jordan, J.; Yousefi, K.; Lam, L.L.; Kolisnik, T.; Chelissery, J.; et al. Low PCA3 expression is a marker of poor differentiation in localized prostate tumors: Exploratory analysis from 12,076 patients. *Oncotarget* **2017**, *8*, 50804–50813. [CrossRef]
42. Lee, G.L.; Dobi, A.; Srivastava, S. Diagnostic performance of the PCA3 urine test. *Nat. Rev. Urol.* **2011**, *8*, 123–124. [CrossRef]
43. Gou, X.; Huang, P.; Mou, C.; Luo, Y. The PCA3 test for guiding repeat biopsy of prostate cancer and its cut-off score: A systematic review and meta-analysis. *Asian J. Androl.* **2014**, *16*, 487–492. [CrossRef] [PubMed]
44. Ramachandran, S.; Liu, P.; Young, A.N.; Yin-Goen, Q.; Lim, S.D.; Laycock, N.; Amin, M.B.; Carney, J.K.; Marshall, F.F.; Petros, A.J.; et al. Loss of HOXC6 expression induces apoptosis in prostate cancer cells. *Oncogene* **2005**, *24*, 188–198. [CrossRef]
45. Zhou, J.; Yang, X.; Song, P.; Wang, H.; Wang, X. *HOXC6* in the prognosis of prostate cancer. *Artif. Cells Nanomed. Biotechnol.* **2019**, *47*, 2715–2720. [CrossRef] [PubMed]
46. Pepe, P.; Dibenedetto, G.; Pepe, L.; Pennisi, M. Multiparametric MRI Versus SelectMDx Accuracy in the Diagnosis of Clinically Significant PCa in Men Enrolled in Active Surveillance. *In Vivo* **2019**, *34*, 393–396. [CrossRef] [PubMed]
47. Hendriks, R.J.; van der Leest, M.M.G.; Israël, B.; Hannink, G.; YantiSetiasti, A.; Cornel, E.B.; de Kaa, C.A.H.-V.; Klaver, O.S.; Sedelaar, J.P.M.; Van Criekinge, W.; et al. Clinical use of the SelectMDx urinary-biomarker test with or without mpMRI in prostate cancer diagnosis: A prospective, multicenter study in biopsy-naïve men. *Prostate Cancer Prostatic Dis.* **2021**, *24*, 1110–1119. [CrossRef] [PubMed]

48. Mitra, S.; Annamalai, L.; Chakraborty, S.; Johnson, K.; Song, X.-H.; Batra, S.K.; Mehta, P.P. Androgen-regulated Formation and Degradation of Gap Junctions in Androgen-responsive Human Prostate Cancer Cells. *Mol. Biol. Cell* **2006**, *17*, 5400–5416. [CrossRef]
49. Ricketts, C.J.; De Cubas, A.A.; Fan, H.; Smith, C.C.; Lang, M.; Reznik, E.; Bowlby, R.; Gibb, E.A.; Akbani, R.; Beroukhim, R.; et al. The Cancer Genome Atlas Comprehensive Molecular Characterization of Renal Cell Carcinoma. *Cell Rep.* **2018**, *23*, 3698. [CrossRef]
50. Erdem, H.; Çırakoğlu, A.; Benli, E.; Çankaya, S. Association of Connexin 32 with Prostate Volume and PSA Level in Prostatic Adenocarcinoma and Adenomyomatous Hyperplasia. *J. Urol. Surg.* **2020**, *7*, 103–108. [CrossRef]
51. Luna-Coronell, J.A.; Vierlinger, K.; Gamperl, M.; Hofbauer, J.; Berger, I.; Weinhäusel, A. The prostate cancer immunome: In silico functional analysis of antigenic proteins from microarray profiling with IgG. *Proteomics* **2016**, *16*, 1204–1214. [CrossRef]
52. Wu, Q. Hepsin and prostate cancer. *Front. Biosci.* **2007**, *12*, 5052–5059. [CrossRef]
53. Xia, C.; Ma, W.; Wang, F.; Hua, S.-B.; Liu, M. Identification of a prostate-specific G-protein coupled receptor in prostate cancer. *Oncogene* **2001**, *20*, 5903–5907. [CrossRef] [PubMed]
54. Han, Z.-D.; Zhang, Y.-Q.; He, H.-C.; Dai, Q.-S.; Qin, G.-Q.; Chen, J.-H.; Cai, C.; Fu, X.; Bi, X.-C.; Zhu, J.-G.; et al. Identification of novel serological tumor markers for human prostate cancer using integrative transcriptome and proteome analysis. *Med. Oncol.* **2012**, *29*, 2877–2888. [CrossRef] [PubMed]
55. Laxman, B.; Morris, D.S.; Yu, J.; Siddiqui, J.; Cao, J.; Mehra, R.; Lonigro, R.J.; Tsodikov, A.; Wei, J.T.; Tomlins, S.A.; et al. A First-Generation Multiplex Biomarker Analysis of Urine for the Early Detection of Prostate Cancer. *Cancer Res* **2008**, *68*, 645–649. [CrossRef]
56. Attard, G.; Clark, J.; Ambroisine, L.; Fisher, G.; Kovacs, G.; Flohr, P.; Berney, D.; Foster, C.S.; Fletcher, A.; Gerald, W.L.; et al. Duplication of the fusion of TMPRSS2 to ERG sequences identifies fatal human prostate cancer. *Oncogene* **2008**, *27*, 253–263. [CrossRef]
57. Clark, J.P.; Cooper, C.S. ETS gene fusions in prostate cancer. *Nat. Rev. Urol.* **2009**, *6*, 429–439. [CrossRef] [PubMed]
58. Clark, J.; Attard, G.; Jhavar, S.; Flohr, P.; Reid, A.; De-Bono, J.; Eeles, R.; Scardino, P.; Cuzick, J.; Fisher, G.; et al. Complex patterns of ETS gene alteration arise during cancer development in the human prostate. *Oncogene* **2007**, *27*, 1993–2003. [CrossRef] [PubMed]
59. Mehra, R.; Han, B.; Tomlins, S.A.; Wang, L.; Menon, A.; Wasco, M.J.; Shen, R.; Montie, J.E.; Chinnaiyan, A.M.; Shah, R.B. Heterogeneity of *TMPRSS2* Gene Rearrangements in Multifocal Prostate Adenocarcinoma: Molecular Evidence for an Independent Group of Diseases. *Cancer Res.* **2007**, *67*, 7991–7995. [CrossRef] [PubMed]
60. Young, A.; Palanisamy, N.; Siddiqui, J.; Wood, D.P.; Wei, J.T.; Chinnaiyan, A.M.; Kunju, L.P.; Tomlins, S.A. Correlation of Urine *TMPRSS2:ERG* and *PCA3* to ERG+ and Total Prostate Cancer Burden. *Am. J. Clin. Pathol.* **2012**, *138*, 685–696. [CrossRef] [PubMed]
61. Tomlins, S.A.; Day, J.R.; Lonigro, R.J.; Hovelson, D.H.; Siddiqui, J.; Kunju, L.P.; Dunn, R.L.; Meyer, S.; Hodge, P.; Groskopf, J.; et al. Urine TMPRSS2:ERG Plus PCA3 for Individualized Prostate Cancer Risk Assessment. *Eur. Urol.* **2015**, *70*, 45–53. [CrossRef]
62. Fujita, K.; Pavlovich, C.P.; Netto, G.J.; Konishi, Y.; Isaacs, W.B.; Ali, S.; De Marzo, A.; Meeker, A.K. Specific detection of prostate cancer cells in urine by multiplex immunofluorescence cytology. *Hum. Pathol.* **2009**, *40*, 924–933. [CrossRef]
63. Quek, S.-I.; Wong, O.M.; Chen, A.; Borges, G.T.; Ellis, W.J.; Salvanha, D.M.; Vêncio, R.Z.; Weaver, B.; Ench, Y.M.; Leach, R.J.; et al. Processing of voided urine for prostate cancer RNA biomarker analysis. *Prostate* **2015**, *75*, 1886–1895. [CrossRef] [PubMed]
64. Leyten, G.H.; Hessels, D.; Smit, F.P.; Jannink, S.A.; de Jong, H.; Melchers, W.J.; Cornel, E.B.; de Reijke, T.M.; Vergunst, H.; Kil, P.; et al. Identification of a Candidate Gene Panel for the Early Diagnosis of Prostate Cancer. *Clin. Cancer Res.* **2015**, *21*, 3061–3070. [CrossRef] [PubMed]
65. Donovan, M.J.; Noerholm, M.; Bentink, S.; Belzer, S.; Skog, J.; O'Neill, V.; Cochran, J.S.; Brown, G.A. A molecular signature of PCA3 and ERG exosomal RNA from non-DRE urine is predictive of initial prostate biopsy result. *Prostate Cancer Prostatic Dis.* **2015**, *18*, 370–375. [CrossRef] [PubMed]
66. Motamedinia, P.; Scott, A.N.; Bate, K.L.; Sadeghi, N.; Salazar, G.; Shapiro, E.; Ahn, J.; Lipsky, M.; Lin, J.; Hruby, G.W.; et al. Urine Exosomes for Non-Invasive Assessment of Gene Expression and Mutations of Prostate Cancer. *PLoS ONE* **2016**, *11*, e0154507. [CrossRef]
67. Ahmed, H.U.; El-Shater Bosaily, A.; Brown, L.C.; Gabe, R.; Kaplan, R.; Parmar, M.K.; Collaco-Moraes, Y.; Ward, K.; Hindley, R.G.; Freeman, A.; et al. Diagnostic accuracy of multi-parametric MRI and TRUS biopsy in prostate cancer (PROMIS): A paired validating confirmatory study. *Lancet* **2017**, *389*, 815–822. [CrossRef]
68. Kim, S.; Schroeder, C.M.; Xie, X.S. Single-Molecule Study of DNA Polymerization Activity of HIV-1 Reverse Transcriptase on DNA Templates. *J. Mol. Biol.* **2010**, *395*, 995–1006. [CrossRef]
69. Bracht, J.W.P.; Gimenez-Capitan, A.; Huang, C.-Y.; Potie, N.; Pedraz-Valdunciel, C.; Warren, S.; Rosell, R.; Molina-Vila, M.A. Analysis of extracellular vesicle mRNA derived from plasma using the nCounter platform. *Sci. Rep.* **2021**, *11*, 3712. [CrossRef]

70. David, A.; Mabjeesh, N.; Azar, I.; Biton, S.; Engel, S.; Bernstein, J.; Romano, J.; Avidor, Y.; Waks, T.; Eshhar, Z.; et al. Unusual Alternative Splicing within the Human Kallikrein Genes KLK2 and KLK3 Gives Rise to Novel Prostate-specific Proteins. *J. Biol. Chem.* **2002**, *277*, 18084–18090. [CrossRef]
71. Hopkins, J.F.; Sabelnykova, V.Y.; Weischenfeldt, J.; Simon, R.; Aguiar, J.A.; Alkallas, R.; Heisler, L.E.; Zhang, J.; Watson, J.D.; Chua, M.L.K.; et al. Mitochondrial mutations drive prostate cancer aggression. *Nat. Commun.* **2017**, *8*, 656. [CrossRef]

*cancers*

MDPI

*Review*

# Bispecific T-Cell Engagers Therapies in Solid Tumors: Focusing on Prostate Cancer

**Diana C. Simão [1], Kevin K. Zarrabi [2], José L. Mendes [1], Ricardo Luz [1], Jorge A. Garcia [3], William K. Kelly [2] and Pedro C. Barata [3,*]**

1 Department of Medical Oncology, Centro Hospitalar Universitário de Lisboa Central, 1169-050 Lisbon, Portugal
2 Department of Medical Oncology, Sidney Kimmel Cancer Center, Thomas Jefferson University, Philadelphia, PA 19107, USA
3 Division of Solid Tumor Oncology, University Hospitals Seidman Cancer Center, Case Western Reserve University, Cleveland, OH 44106, USA
* Correspondence: pedro.barata@uhhospitals.org; Tel.: +1-216-262-1214

**Simple Summary:** Cancer treatments have significantly changed with the introduction of immunotherapy. Recently, the development of new agents that harness the redirection of T-cells against cancer is rapidly emerging in multiple tumor types. Since bispecific T-cell engager (BiTE) therapies have demonstrated clinical benefit in hematologic malignancies, their application to solid tumors has been an active area of investigation. However, in prostate cancer, due to the heterogeneous and immune-suppressive tumor microenvironment, the development of immunotherapy strategies remains a therapeutic challenge. In this review, we summarize the current development of BiTE therapies in solid tumors with a particular focus on clinical trials in advanced prostate cancer.

**Abstract:** Over the past decade, immunotherapy has demonstrated an impressive improvement in treatment outcomes for multiple cancers. Following the landmark approvals for use of immune checkpoint inhibitors, new challenges emerged in various clinical settings. Not all tumor types harbor immunogenic characteristics capable of triggering responses. Similarly, many tumors' immune microenvironment allows them to become evasive, leading to resistance and, thus, limiting the durability of responses. To overcome this limitation, new T-cell redirecting strategies such as bispecific T-cell engager (BiTE) have become attractive and promising immunotherapies. Our review provides a comprehensive perspective of the current evidence of BiTE therapies in solid tumors. Considering that immunotherapy has shown modest results in advanced prostate cancer to date, we review the biologic rationale and promising results of BiTE therapy in this clinical setting and discuss potential tumor-associated antigens that may be integrated into BiTE construct designs. Our review also aims to evaluate the advances of BiTE therapies in prostate cancer, illustrate the major obstacles and underlying limitations, and discuss directions for future research.

**Keywords:** prostate cancer; immunotherapy; bispecific T-cell engagers; prostate-specific membrane antigen; prostate stem cell antigen

## 1. Introduction

Cancer immunotherapy, defined as the science of modulating the immune system against cancer, represents a new paradigm shift in the field of oncology, prolonging survival in several solid tumors [1]. Nevertheless, it took decades of basic science discoveries to demonstrate the ability of modulating the immune system to treat cancer and subsequently establish its role in clinical practice [2].

Since immune response against cancer involves complex interactions between tumor, host, and environment, different strategies have been developed including immunostimulatory cytokines, vaccines, adoptive cell therapies, oncolytic viruses, and immune checkpoint inhibitors (ICIs) [3]. High-dose interleukin-2 (IL-2), a potent inducer of cytotoxic T-cells and NK cells, was one of the first FDA-approved immunotherapy drugs in advanced cancer, with a role in the treatment of melanoma and renal cell carcinoma [4]. Although high-dose cytokine therapy is no longer used due to its short half-life and significant toxicity in therapeutic doses, the activity of IL-2 provided a fundamental understanding of the therapeutic potential of T-cell regulation in the development of new immunotherapy strategies [5].

ICIs targeting CTLA-4 and PD1/PD-L1 pathways emerged as a revolutionary cancer treatment strategy, due to impressive clinical responses and overall outcome benefits in certain tumor types [6,7]. However, the efficacy of ICIs in prostate cancer have been modest, except for mismatch-repair-deficient or microsatellite-instability-high tumors, in which pembrolizumab has been approved in a tumor-agnostic manner [8]. Moreover, the development of therapeutic cancer vaccines led to the approval of sipuleucel-T in metastatic prostate cancer. Despite its proven benefit in overall survival [9], questions remain about its true clinical benefit and its role within the treatment paradigm of metastatic prostate cancer.

Recently, novel immunotherapies that redirect T-cells against tumor antigens through antibody fragments independent of major histocompatibility complex (MHC) presentation have been under investigation. In particular, chimeric antigen-receptor-modified (CAR) T-cells and bispecific T-cell engagers (BiTEs) have demonstrated remarkable clinical responses in hematologic malignancies [10,11].

Nonetheless, the development of these novel T-cell redirection approaches in solid tumors still presents major obstacles that limits its clinical application, including tumor heterogeneity and off-tumor toxicity [12,13].

This comprehensive review aims to describe the current evidence of BiTE therapies in solid tumors with a focus on ongoing clinical trials in the treatment of advanced prostate cancer.

## 2. Bispecific T-Cell Engagers (BiTEs) in Cancer Treatment

Bispecific antibodies (bsAb) were first described by Nisonoff and colleagues in the 1960s, as an antibody-based molecule with two distinct antigen-binding sites, which can function to physically bridge two different cells [14]. By simultaneously binding an antigen on tumor cells and a surface molecule on T-cells, bsAbs can redirect and activate T-cells to induce tumor lysis [15,16].

Since the 1980s, multiple bsAb formats have been developed [15]. In preclinical models, these early constructs showed relatively limited efficacy, with high drug concentrations, high effector-to-target ratios required to induce T-cell-mediated cancer cell lysis, and showed significant "off-target" toxicity [17]. However, novel strategies are evolving to overcome these limitations in order to expand and further optimize bsAb formats.

BsAbs are categorized based on their structure and mechanisms of action, specificity, and affinity for target antigen [18]. From a structural point of view, bsAbs are classified by the presence/absence of the Fc region [19]. BiTEs represent prototypical Fc-free bsAbs, with several new constructs currently under clinical evaluation in solid tumors.

### 2.1. BiTE Design and Mechanism of Action

BiTEs are recombinant proteins composed of two different single-chain variable fragment (scFv) regions from two different monoclonal antibodies. The scFv constructs are covalently connected by a flexible small peptide linker, altogether comprising a ~55 kDa polypeptide chain [17,19]. One scFv-binding domain is engineered to target a select tumor-associated antigen (TAA) and the other scFv domain is typically specific for CD3, the invariant component of the T-cell receptor (TCR) complex [20–22]. The length of the inter-scFv linker varies depending on construct, but linker size does not directly impact tumor-killing activity [23]. However, linker length must account for flexibility, stability, and the orientation of binding interaction between TAA and epitope [19,24].

In general, the TCR complex on the surface of T-cells recognizes antigens that are presented via the major histocompatibility complex (MHC) and this interaction triggers a signaling cascade, involving transcription factor activation and cytoskeletal remodeling, resulting in T-cell activation [20]. BiTEs are unique in their ability to redirect T-cells against TAAs on tumor cells and directly activate T-cells, independent of TCR/MHC interaction [21]. More specifically, T-cell activation by interaction with the TCR complex engages T-cells to form an immune synapse on the surface of tumor cells, resulting in release of cytokines, perforins, and granzymes that induces cancer cell apoptosis [21,25]. Moreover, the activation of effector T-cells occurs only when both scFv-binding domains are engaged with their respective targets [26].

In comparison with alternative bsAbs formats and monoclonal IgG antibodies, BiTEs have a 100-to-10,000-fold higher efficacy in tumor cell lysis with a low ratio of T-cells to target tumor cells in cellular models [27]. Subsequent to BiTE-induced T-cell activation, the diffusion of released cytokines in the immune synapse also plays a role in upregulation of cell surface molecules of the surrounding cells, further contributing to the antitumor activity of BiTEs, commonly named the "bystander effect" [28].

Furthermore, BiTEs can be produced in large quantities by mammalian cell lines, minimizing interpatient variability and providing "off-the-shelf" therapies that now are undergoing investigation in a multitude of tumor types [29].

### 2.2. Blinatumomab, the First BiTE Construct in Clinical Practice

The CD19/CD3 BiTE molecule blinatumomab has served as clear proof of concept of antitumor activity and clinical efficacy of T-cell engagers in B-cell malignancies. Blinatumomab's clinical efficacy and favorable safety profile lead to its first-in-class approval by the European Medicines Agency (EMA) and the U.S. Food and Drug Administration (FDA) for the treatment of both children and adults with relapsed or refractory Philadelphia chromosome (Ph)-negative precursor B-cell acute lymphoblastic leukemia (B-ALL) [30,31]. In the TOWER trial, a multicenter, international, phase 3 clinical trial, blinatumomab demonstrated a significant improvement in overall survival (7.7 months vs. 4.0 months; HR 0.71; 95% CI 0.55–0.93) and higher rates of hematological remission, compared to standard-of-care chemotherapy [32]. The FDA also approved blinatumomab for relapsed or refractory Ph-positive B-ALL, based on the ALCANTARA trial results that demonstrated remarkable long-term durability of responses in this setting [33]. Long-term follow-up data show that Blinatumomab treatment can render a complete response with minimal residual disease (MRD) in approximately three-quarters of treated patients [33,34], resulting in approval for treatment of MRD-positive patients with B-ALL.

Despite its efficacy, blinatumomab has also been associated with significant adverse events including cytokine release syndrome (CRS) and neurotoxicity, which may be life-threatening [35]. Treatment of patients requires inpatient hospitalization at time of drug administration for monitoring and to allow for prompt management of these potential events.

### 2.3. BiTE Therapy Safety Considerations: Cytokine Release Syndrome and Neurotoxicity

As the first marketed therapy within its class, blinatumomab's safety profile and dose-limiting toxicities are well-described [35]. The two main toxicity concerns associated with BiTE immunotherapy correlate with its mechanism of action and include cytokine release syndrome (CRS) and neurotoxicity.

CRS is an uncontrolled systemic inflammatory response characterized by high levels of pro-inflammatory cytokines, most notably interferon (IFN)-gamma, IL-1, and IL-6, and is induced by T-cell activation [36]. CRS clinical manifestations and severity varies from mild fever or rash to severe multi-organ failure [36]. When CRS occurs after dose administration of blinatumomab, symptoms usually appear during the first infusion cycle but can be delayed by days, and the risk of grade $\geq$3 CRS ranges from 2 to 11% for B-cell malignancies [32–35]. A higher incidence of CRS has been associated with higher tumor burden and drug dosage [36,37]. Early intervention is critical to prevent progression to life-threatening toxicity. In patients with mild CRS, supportive care is indicated, while grade $\geq$3 CRS is managed with infusion interruption and immunosuppression with glucocorticoids [38].

Tocilizumab, a humanized monoclonal antibody that targets the IL-6 receptor, was approved for the management of severe or life-threatening CRS [38]. To reduce the incidence of CRS, prophylactic use of dexamethasone combined with step-dosing administration of blinatumomab is recommended [36,37].

Similar to CRS, immune effector-cell-associated neurotoxicity syndrome (ICANS) pathophysiology is complex and incompletely understood and seems to be related to the production of pro-inflammatory cytokines, with subsequent T-cells adhesion to brain endothelium and disruption of the blood–brain barrier [39,40]. The incidence of grade $\geq$3 neurotoxicity with blinatumomab ranges from 5.5 to 24% [31–34]. Neurotoxicity occurs most commonly in the first treatment cycle and risk increases with higher doses of blinatumomab. The most common symptoms include dizziness, tremor, confusional state, and encephalopathy [41]. Management of ICANS require treatment interruption and corticosteroids; however, definitive evidence is lacking as to whether corticosteroids have a beneficial effect on the severity or duration of ICANS [39].

In addition, other relevant adverse events related to the CD19-targeting mechanism have been reported with blinatumomab, namely long-term B-cell aplasia and hypogammaglobulinemia [33–35]. As such, immunoglobulin replacement and prophylactic antibiotics should be considered on an individual case-by-case basis.

### *2.4. Limitations of BiTE Therapies and Innovative Strategies to Enhance Efficacy*

A practical limitation of the prototypical BiTE molecule in clinical practice is their short half-life and the need for continuous intravenous administration [17]. New approaches to optimize drug delivery and alter pharmacokinetics include half-life-extended (HLE) BiTEs [42,43], which are single-chain polypeptides incorporating an additional Fc region, creating a bsAb with a higher molecular weight and extended half-life. With regards to alternate routes of administration, subcutaneous BiTEs have been investigated, showing a manageable safety profile similar to that previously reported for intravenous formulations [44].

"Off-the-shelf" BiTE manufacturing is a major advantage of this treatment modality, supporting its clinical applicability and cost-effectiveness, since large quantities are produced, without interpatient variability [29]. However, identification of target antigens that are ubiquitously expressed on tumor cells in all patients has been a critical challenge for the application of BiTE therapy to certain cancers, particularly solid malignancies. Many tumor-specific antigens are intracellular and are not accessible for standard T-cell engagers, while numerous cell-surface TAAs overexpressed in solid tumors lack high specificity and are often found at low levels in normal tissue [21,22]. As a consequence, "on-target, off-tumor" toxicity has been a challenge in TAA selection for solid tumors [45]. The balance between maximizing therapeutic potential of BiTEs while mitigating toxicity remains an area which requires further investigation [42].

Acquired treatment resistance to BiTE therapy is yet another limitation. Downregulation or loss of TAA expression has been described as a major mechanism of resistance to BiTE therapies [46]. This observation gave rise to the development of multiple novel T-cell engager constructs, with different pharmacokinetic and pharmacodynamic profiles, including different formats with higher stability in a construct that enables optimal interaction between the target and effector cell (dual affinity retargeting (DART®) bispecific antibodies), and some simultaneously targeting different TAAs (simultaneous multiple interaction T-cell engagers (SMITEs)) [17].

Another mechanism of resistance to BiTE therapy is the upregulation of inhibitory immune checkpoints within the tumor microenvironment (TME) [47]. The TME of solid tumors contains a complex composition of malignant, immune, and stromal cell populations that can suppress antitumor T-cell responses, which negatively affects T-cell engager efficacy [48]. To overcome this limitation, constructs with concomitant immune-checkpoint action (checkpoint inhibitor T cell-engagers) that target the PD-1/PD-L1 axis are under development, as well as combination strategies combining BiTEs with ICIs [49].

## 3. BiTEs in Solid Tumors

Despite the impressive results of BiTEs in hematological malignancies, illustrated by blinatumomab efficacy in B-ALL, first-generation compounds have failed to demonstrate significant antitumor activity in solid tumors [13,50]. Catumaxomab was the first bispecific T-cell engager approved by the EMA in 2009 to treat malignant ascites of epithelial cancers [51]. It is a trifunctional bispecific IgG antibody, with one arm recognizing the epithelial cell adhesion molecule (EpCAM) on tumoral cells and another arm targeting the CD3 subunit on T-cells. Furthermore, the functional Fc fragment binds to different immune accessory cells, such as monocytes, macrophages, and natural killer (NK) cells, inducing T-cell activation and NK cell recruitment [52]. EMA approval was based on an improvement of puncture-free survival and signs and symptoms of ascites from a large, randomized phase 2/3 trial [51]. However, intravenous administration of catumaxomab was associated with severe adverse events, including CRS and dose-dependent liver toxicity [53], with one patient experiencing fulminant fatal acute liver failure that led to the early termination of the trial, and later withdrawal of catumaxomab from the market.

Solitomab is another first-generation BiTE targeting EpCAMxCD3 (MT110 or AMG110) that was investigated in a phase 1 trial of 65 patients with relapsed/refractory advanced-stage solid cancers. Treatment was associated with dose-limiting toxicities, including severe diarrhea and increased liver enzymes, which precluded dose escalation to potentially therapeutic levels [54].

Subsequent next-generation BiTE molecules have been constructed which are directed against TAAs with reduced expression in non-neoplastic tissue and employ formats that do not include an Fc domain.

Specifically, in CEA-positive solid tumors such as metastatic colorectal cancer, RO6958688 (also known as CEA CD3 T-cell bispecific or RG7802) was administered as monotherapy or in combination with atezolizumab in a phase 1 trial that enrolled 35 patients. Antitumor activity was observed in monotherapy, with two patients showing partial response, which was enhanced when in combination with atezolizumab, and with a manageable safety profile [55].

More recently, preliminary data from the DUET-1 phase 1 trial showed that tidutamab (XmAb18087), a BiTE-targeting somatostatin receptor 2 (SSTR2), was well tolerated with a best overall response of stable disease in patients with advanced neuroendocrine tumors [56].

We are now witnessing increasing numbers of bispecific-based T-cell engagers undergoing rapid development and evaluation in several tumor types (Table 1). Prostate cancer target-antigens and clinical trials will be further discussed separately.

**Table 1.** Bispecific T-cell engagers therapy in solid tumors (clinicaltrials.gov accessed on 1 November 2022).

| NCT | Phase | Drug (Format) | Target | Indication | Status | Results | Ref. |
|---|---|---|---|---|---|---|---|
| | | | | EpCAM | | | |
| NCT00836654 | 2/3 | Catumaxomab or Removab® (Triomab®) | EpCAM xCD3 | Malignant ascites and EpCAM-positive tumors | Completed | N = 258 (129 ovarian cancer)<br>Puncture-free survival: 46 vs. 11 days<br>AE: fever (60%); abdominal pain (43%) | [51] |
| NCT01065246 | 2 | Catumaxomab | EpCAM xCD3 | Malignant ascites due to epithelial carcinoma | Completed | N = 8 (rechallenge of intraperitoneal catumaxomab)<br>Puncture-free survival: 47.5 days | [57] |
| NCT00326885 | 2 | Catumaxomab | EpCAM xCD3 | Malignant ascites ovarian cancer | Completed | N = 32<br>Puncture-free survival: 29.5 days<br>Ascites symptoms improved | [58] |
| NCT01246440 | 2 | Catumaxomab | EpCAM xCD3 | Ovarian cancer | Completed | N = 46 (consolidation therapy)<br>Median duration treatment: 13 days<br>Grade 3–4 AE in 29 pts (74.4%)<br>Treatment interruption in 4 (10.2%) | [59] |
| NCT00189345 | 2 | Catumaxomab | EpCAM xCD3 | Platinum refractory ovarian, fallopian tube, and peritoneal neoplasms | Completed | N = 46<br>(low dose 23 + high dose 22)<br>No difference AE low vs. high<br>Stable disease in 2 pts (low) and 5 pts (high) | [60] |
| NCT01815528 | 2 | Catumaxomab | EpCAM xCD3 | Recurrent epithelial ovarian cancer | Completed | Not reported | |
| NCT00563836 | 2 | Catumaxomab | EpCAM xCD3 | Ovarian cancer | Completed | Not reported | |
| NCT04222114 | 3 | Catumaxomab | EpCAM xCD3 | Gastric cancer | Recruiting | | |

**Table 1.** *Cont.*

| NCT | Phase | Drug (Format) | Target | Indication | Status | Results | Ref. |
|---|---|---|---|---|---|---|---|
| NCT01504256 | 2 | Catumaxomab + FLOT | EpCAM xCD3 | Gastric adenocarcinoma with peritoneal carcinomatosis | Completed | N = 31 (FLOT + catumaxomab 15 pts (A) vs. FLOT alone 16 pts (B)) Complete remission of carcinomatosis: 27% (A) vs. 19% (B) ($p = 0.69$). Severe AE: fever (23%), abdominal pain (31%), elevated liver enzymes (31%). Median PFS: 6.7 (A) vs. 5.4 months (B) ($p = 0.71$). | [61] |
| NCT00464893 | 2 | Catumaxomab | EpCAM xCD3 | Gastric cancer | Completed | Not reported | |
| NCT00352833 | 2 | Catumaxomab | EpCAM xCD3 | Gastric cancer | Completed | Not reported | |
| NCT04501744 | 1 | M701 | EpCAM xCD3 | Malignant ascites | Recruiting | | |
| NCT00635596 | 1 | Solitomab or MT110 or AMG110 | EpCAM xCD3 | Relapsed/refractory solid tumors | Completed | N = 65 (35 colorectal; 10 ovarian; 8 gastric; 6 NSCLC; 3 SCLC; 3 mCRPC) 95% Grade ≥ 3 AE, mainly diarrhea, elevated liver parameters and lipase | [54] |
| | | | | CEA | | | |
| NCT02324257 NCT02650713 | 1 | RO6958688 or RG7802 + atezolizumab | CEA xCD3 | CEA-positive tumors | Completed | N = 36 pts in monotherapy + 10 pts in combination Grade ≥ 3 AEs: infusion related reaction (16.3%) and diarrhea (5%) | [55] |
| NCT01284231 | 1 | AMG211 or MEDI-565 | CEA xCD3 | Gastrointestinal adenocarcinomas | Completed | N = 39 (28 colorectal, 6 pancreatic, 5 other) Grade ≥ 3 AE in 5 pts (hypoxia n = 2, diarrhea, and CRS) Stable disease in 11 pts (28%) | [62] |
| NCT02291614 | 1 | AMG211 or MEDI-565 | CEA xCD3 | Gastrointestinal adenocarcinomas | Completed | Terminated due to high immunogenicity at high doses of >3.2 mg | [63] |
| NCT03337698 | 1/2 | RO6958688 + atezolizumab | CEA xCD3 | NSCLC | Recruiting | | [64] |

**Table 1.** *Cont.*

| NCT | Phase | Drug (Format) | Target | Indication | Status | Results | Ref. |
|---|---|---|---|---|---|---|---|
| | | | | EGFR | | | |
| NCT02620865 | 1/2 | EGFR Bi-armed activated T-cells (BATs) | EGFR xCD3 | Advanced pancreatic cancer | Completed | N = 7<br>No dose-limiting toxicities (DLTs),<br>Median time to progression: 7 months | [65] |
| NCT03269526 | 1/2 | EGFR BATs | EGFR xCD3 | Advanced pancreatic cancer | Recruiting | | |
| NCT03296696 | 1 | AMG596 | EGFRvIII xCD3 | Glioblastoma | Completed | Not reported | [66] |
| NCT03344250 | 1 | EGFR BATs + Temozolomide+ RT | EGFR xCD3 | Glioblastoma | Active, not recruiting | | |
| | | | | gpA33 | | | |
| NCT02248805 | 1 | MGD007 (DART®) | gpA33 xCD3 | Metastatic CRC | Completed | Not reported | [67] |
| NCT03531632 | 1/2 | MGD007 + MGA012 | gpA33 xCD3 | Metastatic CRC | Completed | Not reported | |
| | | | | HER2 | | | |
| NCT04501770 | 1 | M802 | HER2 xCD3 | HER2-positive advanced solid tumors | Not yet recruiting | | |
| NCT03448042 | 1 | Runimotamab + trastuzumab + tocilizumab | HER2 xCD3 | Locally advanced or metastatic HER2-expressing solid tumors | Recruiting | | |
| NCT03272334 | 1/2 | HER2 BATs + Pembrolizumab | HER2 xCD3 | Metastatic breast cancer | Recruiting | | |

**Table 1.** *Cont.*

| NCT | Phase | Drug (Format) | Target | Indication | Status | Results | Ref. |
|---|---|---|---|---|---|---|---|
| | | | | Other | | | |
| NCT03411915 | 1 | Tidutamab (XmAb18087) | SSTR2 xCD3 | NET and GIST | Completed | N = 41<br>Grade $\geq$ 3 AE: lymphopenia (29.3%): transaminase and GGT increase (19.5%); hypophosphatemia (9.8%) and lipase increase (7.3%) | [56] |
| NCT04590781 | 1/2 | Tidutamab (XmAb18087) + Pembrolizumab | SSTR2 xCD3 | Advanced Merkel cell carcinoma and ES-SCLC | Completed | Not reported | |
| NCT04424641 | 1/2 | GEN1044 (DuoBody®) | 5T4 xCD3 | Solid tumors | Completed | Results on submission clinicaltrials.gov (accessed on 1 November 2022) | |
| NCT05180474 | 1 | GEN1047 (DuoBody®) | B7H4 xCD3 | Solid tumors | Recruiting | | |
| NCT04083599 | 1/2 | GEN1042 | 4-1BB xCD40 | Solid tumors | Recruiting | | |
| NCT04496674 | 1 | CC-1 + Tocilizumab | PSMA xCD3 | NSCLC | Recruiting | | |
| NCT04260191 | 1 | AMG910 | CLDN18.2 xCD3 | Gastric and gastroesophageal junction adenocarcinoma | Active, not recruiting | | |
| NCT03146637 | 2 | Activated CIK | MUC1/CEA/ Ep-CAM/GPC3xCD3 | Advanced liver cancer | Recruiting | | |
| NCT03319940 | 1 | AMG757 (HLE) + Pembrolizumab | DLL3 | SCLC | Recruiting | | [68] |

**Table 1.** *Cont.*

| NCT | Phase | Drug (Format) | Target | Indication | Status | Results | Ref. |
|---|---|---|---|---|---|---|---|
| NCT04471727 | 1/2 | HPN328 (TriTAC) | DLL3 | SCLC | Recruiting | | |
| NCT04590326 | 1/2 | REGN4018 or REGN5668 + Cemiplimab | MUC16 xCD3 or MUC16 xCD28 | Ovarian cancer, fallopian tube cancer, peritoneal cancer | Recruiting | | [69] |
| NCT03564340 | 1/2 | REGN4018 + Cemiplimab | MUC16 xCD3 | Ovarian cancer, fallopian tube cancer, peritoneal cancer | Recruiting | | |
| NCT04117958 | 1 | AMG199 (HLE) | MUC17 xCD3 | MUC17-positive solid tumors | Recruiting | | |

Recently, the FDA and EMA approved a bispecific fusion protein indicated for the treatment of adult patients with HLA-A*02:01-positive metastatic uveal melanoma [70,71]. Tebentafusp is a first-in-class immune-mobilizing monoclonal T cell receptor (TCR) against cancer (ImmTAC®), comprising a TCR domain that binds with high affinity to a gp100 peptide presented by human leukocyte antigen—A*02:01 (HLA-A*02:01) on the cell surface of uveal melanoma tumor cells, and an effector domain which binds to the CD3 receptor on polyclonal T-cells. Tebentafusp significantly improved overall survival in patients with previously untreated metastatic uveal melanoma in a large, randomized, phase 3 study that led to its regulatory approval [72].

## 4. BiTEs in Advanced Prostate Cancer

The success of immunotherapy in treatment of advanced prostate cancer has been modest, as most modern immunotherapies have failed to achieve long-term remissions. To date, sipuleucel-T is the only approved immunotherapy for metastatic prostate cancer, yet it is not considered a cornerstone therapy for men with metastatic castration-resistant prostate cancer (mCRPC). Sipuleucel-T incorporates autologous antigen-presenting cells (APCs), with a recombinant fusion protein (PA2024), consisting of a prostate antigen, prostatic acid phosphatase (PAP), and granulocyte–macrophage colony-stimulating factor (GMCSF). Although survival benefit was demonstrated in patients with mCRPC [9], questions about the true mechanism of action of this agent still remain [73], thus limiting its application in daily clinical practice.

Except for tumors associated with microsatellite instability, most mCRPC tumors are considered immunologically "cold", due to lack of pro-inflammatory cytokine production, sparse T-cell infiltration, and predominance of suppressive immune components [74,75]. To characterize tumor microenvironment of bone metastasis of prostate cancer, Kfoury et al. performed single-cell analysis and found bone marrow infiltration of tumor-associated macrophages and monocytes with overexpression of cytokine CCL2, leading to T-cell exhaustion as a mechanism of immunosuppression [76].

Furthermore, recent findings described T-cell-intrinsic androgen activity as a novel mechanism of resistance to immunotherapy [77].

Novel immunotherapies using MHC-independent T-cell redirection and activation have been an active area of research with hopes to overcome the immunosuppressive TME within prostate cancer [78]. Particularly in mCRPC, tumor-specific surface markers with relatively low expression in normal tissues have been investigated as potential TAA targets of novel T-cell redirection strategies. Currently, there are several targets for BiTE therapies under development in the prostate cancer disease space (Table 2), including constructs targeting prostate-specific membrane antigen (PSMA), prostate stem cell antigen (PSCA), six-transmembrane epithelial antigen of the prostate (STEAP-1), and Notch ligand delta-like protein 3 (DLL3).

**Table 2.** Clinical trials of Bispecific T-cell engagers in prostate cancer. (clinicaltrials.gov accessed on 10 November 2022).

| NCT | Phase | Drug (Format) | Target | Indication | Status | Results | Ref. |
|---|---|---|---|---|---|---|---|
| NCT01723475 | 1 | Pasotuxizumab, BAY2010112 or AMG212 | PSMA xCD3 | mCRPC | Completed | N = 47 (31 sc + 16 iv)<br>AE Grade 3: 53%<br>MTD not reached due to early stop<br>>50% PSA decline in 9 sc + 3 iv pts | [79,80] |
| NCT03792841 | 1 | Acapatamab or AMG160 (HLE) + Pembrolizumab | PSMA xCD3 | mCRPC | Active, not recruiting | N = 43 (monotherapy)<br>Grade 3 CRS: 25.6%<br>MTD not yet reached<br>>50% PSA decline in 12/35 (34.3%) | [81,82] |
| NCT03792841 | 1 | Acapatamab + Enzalutamide + Abiraterone + AMG 404 | PSMA xCD3 | mCRPC | Active, not recruiting | | [83] |
| NCT02262910 | 1 | ES414 or MOR209 or APVO411 (ADAPTIR®) | PSMA xCD3 | mCRPC | Completed | Discontinued due to high immunogenicity of the construct | [84] |
| NCT03577028 | 1/2 | HPN424 (TriTAC®) | PSMA xCD3 | mCRPC | Unknown | N = 80<br>Grade 3 CRS: 4%<br>MTD not yet reached<br>PSA decline in 13/63 pts (21%), including 3 PSA50, 2 PSA30 responses. | [85] |
| NCT03926013 | 1 | JNJ-63898081 or JNJ-081 (DuoBody®) | PSMA xCD3 | mCRPC | Completed | N = 39 (27 sc + 12 iv)<br>All pts $\geq$ 1 treatment-emergent AE<br>No grade $\geq$ 3 CRS<br>>50% PSA decline in 2 pts | [86] |
| NCT04104607 | 1 | CC-1 (IgGsc) | PSMA xCD3 | mCRPC | Recruiting | | [87] |
| NCT04077021 | 1 | CCW702 (DUPA) | PSMA xCD3 | mCRPC | Recruiting | | [88] |
| NCT05125016 | 1/2 | REGN4336 + cemiplimab | PSMA xCD3 | mCRPC | Recruiting | | [89] |

**Table 2.** *Cont.*

| NCT | Phase | Drug (Format) | Target | Indication | Status | Results | Ref. |
|---|---|---|---|---|---|---|---|
| NCT04740034 | 1 | AMG340 or TNB-585 | PSMA xCD3 | mCRPC | Recruiting | | [90] |
| NCT05369000 | 1 | LAVA-1207 (Gammabody®) | PSMA xV$\gamma$9V$\delta$2 | mCRPC | Recruiting | | |
| NCT03927573 | 1 | GEM3PSCA | PSCA xCD3 | PC, NSCLC, Renal cancer | Recruiting | | |
| NCT04221542 | 1 | AMG 509 (XmAb®) + Enzalutamide + Abiraterone | STEAP1 xCD3 | mCRPC | Recruiting | | [91] |
| NCT04702737 | 1b | Tarlatamab or AMG757 (HLE) | DLL3 | NEPC | Recruiting | | [92] |
| NCT03406858 | 2 | HER2Bi-armed activated T cells + Pembrolizumab | HER2B xCD3 | mCRPC | Recruiting | | [93] |

### *4.1. BiTEs Targeting PSMA*

PSMA is a type II transmembrane protein that is highly expressed on the surface of malignant prostate tissue, with variable low expression in non-neoplastic prostate tissue. Non-malignant tissues expressing PSMA include kidney proximal tubules, salivary glandular cells, and the gastrointestinal tract [94,95]. Additionally, PSMA is also expressed on the tumor-associated neovasculature of different solid malignancies [96,97], increasing the research focus on PSMA as TAA in non-prostate cancer therapies.

In addition, PSMA has been widely explored as a biomarker of prostate cancer activity for disease imaging, using radioactive PSMA tracers [98,99], and as a disease-specific target in the field of PSMA theranostics [100,101].

#### 4.1.1. Pasotuxizumab (BAY2010112 or AMG 212)

The first prototypical PSMA-targeting BiTE was pasotuxizumab (BAY2010112 or AMG 212). In a phase 1 clinical trial (NCT01723475), safety and maximum tolerated dose (MTD) of pasotuxizumab in mCRPC was evaluated. As part of an interim data monitoring analysis, it was reported that all 31 patients receiving subcutaneous (SC) injection had developed antidrug antibodies (ADAs) and prophylactic dexamethasone had no effect in mitigating this event [79]. The SC route of administration was discontinued, and the study proceeded with the continuous intravenous (cIV) infusion cohort only [80]. Hummel et al. reported a >50% PSA reduction in nine and three patients in the SC and cIV cohorts, respectively, including two long-term responders [79,80]. No grade 5 adverse events (AE) were reported. Nevertheless, all patients had at least one AE, including fever (85%), chills (38%), and fatigue (34%) [79]. MTD was not reached due to premature termination of the trial in favor clinical study of a next-generation BiTE construct targeting PSMA, AMG-160.

#### 4.1.2. Acapatamab (AMG-160)

Preclinical studies with a half-life extended BiTE, acapatamab (formerly AMG-160), demonstrated T-cell activation in human samples and upregulation of PD-L1 [102,103]. Given the interest in the development of next-generation PSMA-targeting BiTEs, acapatamab is further being assessed in early-phase clinical study. In an ongoing phase 1 clinical trial (NCT03792841) enrolling 43 patients with mCRPC, dose exploration of acapatamab is being evaluated in monotherapy and in combination with pembrolizumab [81,82]. Preliminary results of monotherapy, in a heavily pre-treated population (with a median of four prior lines of therapy), showed PSA reductions >50% occurred in 12/35 (34.3%) evaluable patients at data cut-off (July 2020) [81]. The majority of patients (n = 41; 95.3%) experienced an any-grade AE, including grade 3 CRS in 11 patients (25.6%). No grade 5 events or treatment discontinuation has been observed, although MTD has not yet been reached [81]. An exploratory phase of this trial assessing the combination of acapatamab with pembrolizumab is in progress.

Alternative combination therapy strategies with acapatamab are also being evaluated. An elegantly designed phase 1/2 trial (NCT04631601) consisting of three subprotocols evaluating safety, tolerability, and MTD of acapatamab, in combination with enzalutamide, abiraterone, or AMG 404, a monoclonal antibody targeting PD-1 receptor, is currently recruiting patients [83].

#### 4.1.3. Novel Emerging Constructs Targeting PSMA

ES414/APVO4141/MOR209 is a bispecific antibody constructed with an ADAPTIR® format, which contains a modified antibody Fc region that improves serum stability but does not cross-link T-cells or target cells and two scFv fragments each targeting PSMA and CD3 [84]. In comparison with prototypical BiTE formats, it has a prolonged half-life; however, high immunogenicity of this construct gave rise to unacceptable systemic toxicity and to early discontinuation of the clinical trial.

In contrast with previous BiTE structural design, HPN424 is a trispecific antibody (TriTAC®) designed with three binding domains. In addition to PSMA and CD3 targeting, it incorporates an albumin-binding domain to prolong serum half-life and increase stability [104]. Data from a phase 1/2a clinical trial (NCT03577028) including men with mCRPC who have received more than two prior systemic therapies, showed that 21% of patients had post-baseline PSA declines and reduction in circulating tumor cells (CTC) occurred in 32 of 56 pts (57%) with measurable CTC at baseline. Most common grade >3 AE were AST increase (18%), anemia (11%), and ALT increase (11%). Of note, grade 3 CRS was observed in 4% of patients, occurring with first dose administration, and MTD was not reached [85].

JNJ-63898081/JN-081 is a novel bispecific antibody targeting PSMA and CD3 engineered by a next-generation technology platform denominated DuoBody®. This technology is believed to produce more stable bsAb constructs and retain endogenous IgG structure and pharmacokinetics possibly leading to improved tolerability and efficacy. In a phase 1 trial, all 39 patients enrolled experienced at least one treatment-emergent AE, most commonly pyrexia (n = 27; 69.2%) and CRS (n = 26; 66.7%). Grade 2 CRS was observed at higher doses and was partially mitigated by SC and step-up dosing, with no reported grade ≥3 CRS. Transient PSA decreases were observed at treatment doses greater than 30 μg/kg SC [86].

CC-1 is a bsAb targeting PSMA with a unique IgG-based structure. CC-1 was developed in an IgGsc format (IgG molecule with two c-terminal single chain moieties), which includes an Fc domain of an IgG1 antibody linked to two scFv-binding domains targeting PSMA and CD3. The modified Fc domain is expected to prolong its serum half-life, as well as lower immunogenicity and thereby mitigate toxicity. Despite preliminary data of 14 patients from a dose-escalation phase 1 trial which showed that the majority of patients suffered from a CRS event (79% of patients), the CRS did not exceed grade 2 and resolved in most cases without the need for administration of tocilizumab. Results from a dose expansion of this phase 1 clinical trial are highly anticipated and the study is currently enrolling patients (NCT04104607) [87].

Additional ongoing trials concerning bispecific antibodies targeting PSMA are summarized in Table 2.

### 4.2. *Other Potential TAAs in Prostate Cancer*

#### 4.2.1. Prostate Stem Cell Antigen (PSCA)

PSCA is a cell-surface glycoprotein encoded by the PSCA gene, which is overexpressed in prostate gland cells as well as urothelial, pancreatic, renal, and non-small cell lung cancer [105,106]. Although its biological function is not completely understood, PSCA has been associated with advanced disease and poor prognosis in prostate cancer [107].

An open-label, dose escalation clinical trial of GEM3PSCA (NCT03927573), a PSCAxCD3 bispecific antibody (ATAC® format) is enrolling patients with PSCA-expressing tumor types after failure of standard therapy.

#### 4.2.2. Six-Transmembrane Epithelial Antigen of the Prostate-1 (STEAP-1)

Acting as a membrane channel or transporter protein in cell junctions of epithelial cells, STEAP-1 is overexpressed on the surface of prostate cancer cells with low or no expression on normal tissue [108].

Binding simultaneously to STEAP-1 and CD3, AMG 509 is a bispecific T-cell engager (XmAb® format) that is being evaluated in a phase 1 trial (NCT04221542) as monotherapy and in combination with enzalutamide or abiraterone [91].

### 4.3. *Neuroendocrine Prostate Cancer*

De novo neuroendocrine prostate cancer (NEPC) is a rare tumor, corresponding to less than 2% of all cancers at the time of diagnosis [109]; however, treatment-related-NEPC is found in 10.5–17% of patients with mCRPC after treatment with androgen signaling inhibitors [110].

The 2022 WHO Classification of Tumors of the Urinary and Male Genital System describes treatment-related neuroendocrine prostatic carcinoma (t-NEPC) as a distinct entity: "tumors demonstrating complete neuroendocrine differentiation or partial neuroendocrine differentiation with adenocarcinoma following androgen deprivation therapy" [111]. Regarding histological and immunological features, some are pure small cell, or less commonly large cell, neuroendocrine carcinoma, while others are mixed tumors with a component of high-grade adenocarcinoma [112].

Although it is unclear whether de novo NEPC and t-NEPC have a shared clonal origin, emerging evidence suggests that adenocarcinoma to NEPC transdifferentiation may be driven by concomitant inactivation of TP53 and RB1, and alterations of epigenetic regulation and transcription factors [113].

Initially identified as a surface protein overexpressed in small cell lung cancer, DLL3 is also highly expressed in neuroendocrine prostate cancer (NEPC) [114], emerging as a potential candidate to target for T-cell redirecting therapies since no standard treatment approach for NEPC exists and it remains an unmet need.

Tarlatamab or AMG 757 is an HLE BiTE (DLL3xCD3) that is being evaluated in a phase 1b trial (NCT04702737), which is recruiting patients with metastatic de novo or treatment-emergent NEPC [87].

Novel TAAs for NEPC have been investigated. Han et al. suggested that KIT pathway inhibition may be a potential target in NEPC treatment [115].

### *4.4. Overcoming Hurdles to Successful Implementation of BiTE Therapy within the mCRPC Treatment Paradigm*

Successful development of BiTEs requires a tumor-restricted TAA which allows for effective antigen selectivity and minimal "off-tumor, on-target" toxicity. Prostate cancer is an ideal disease setting for BiTE development given the myriad of unique TAAs that have been discovered. Initial issues with drug immunogenicity and rapid drug clearance have been addressed with the advent of HLE BiTEs. Although toxicity in the early-generation prostate cancer BiTE trials has been ubiquitous, it has also been manageable. Our understanding of BiTE toxicity is informed from prior experiences with hematologic malignancies, and early intervention at first sign of CRS with systemic corticosteroids and immunosuppression has proven to be an effective approach at safely achieving greater therapeutic dose thresholds for optimal drug delivery.

As we optimize the BiTE format in respect to pharmacokinetics, tissue selectivity, and toxicity, we are also in need of TME modulation to allow BiTE effector function within a less immunologically 'cold' and suppressive TME. Multimodal therapies with BiTEs in combination with novel modalities are a strategy under investigation. We now have a clinical study employing BiTEs in combination with ICIs in prostate cancer, which is a strategy with strong biologic rationale as disinhibiting immune effector cells can help augment BiTE response.

The trials to date in prostate cancer employ a heavily pre-treated patient population, as is the case of BiTE trials in other solid-tumor malignancies. In patients with mCRPC, prior treatment with anti-androgen therapy, theranostics therapy, and typically multiple lines of cytotoxic chemotherapy may not be an ideal sequence prior to T-cell immunotherapy. Development of castration resistance after anti-androgen therapy, as well as usage of cytotoxic chemotherapy, each accelerate development of an exhausted T-cell phenotype rich in T-regulatory cells, which is not conducive for immune response. As such, earlier integration of T-cell immunotherapy and early referral for trial consideration is recommended.

## 5. Future Perspectives

The underlying causes of the limited effectiveness of BiTE therapy in solid-tumor malignancies are multifactorial. It is unlikely that a single strategy to optimize BiTE construct design will dramatically improve treatment efficacy. Ongoing clinical investigation into the immune-escape mechanisms and better characterization of the immune milieu within the

TME of each respective tumor type will be critical in order to identify novel target antigens which can then be integrated into next-generation BiTE construct design. Moreover, the success of BiTE therapy in solid-tumor malignancies will likely rely on two key additional factors: (1) mitigation of severe toxicity and (2) ideal identification and selection of patients who may benefit most from immunologic response.

BiTE therapy toxicity, namely CRS, is highly predictable and some form of systemic inflammatory response is expected in nearly all patients. Early identification of toxicity and intervention with corticosteroids has certainly proven efficacious in preventing severe toxicity leading to end-organ dysfunction. There is ongoing concern that immunosuppressive intervention for CRS will counteract any anti-tumor immune response triggered by therapy. However, recent studies evaluating early administration of tocilizumab with a newly developed PSMAxCD3 bsAb revealed reduction in undesired sequelae of CRS without affecting therapeutic activity [116]. Taking early measures to attenuate immune-related toxicity may prevent dose-reduction and early drug discontinuation, thereby allowing patients to benefit from more drug exposure and maximized treatment effect.

Ideal patient selection represents yet another challenge which may be an area of focus for future research. We now have identified a plethora of tumor-associated antigens (TAAs) that demonstrate high specificity for target antigen-expressing tumor, yet TAA as a biomarker has been insufficient in patient selection to identify exceptional responders. It is very possible that chemotherapy refractory and heavily pre-treated patient populations preclude optimal patient selection. The timing of BiTE therapy within the treatment cascade also requires further investigation and is an ongoing area of research [117]. Early introduction of BiTE therapy within the disease course may be a pathway to maximize therapeutic potential prior to development of an exhausted immune phenotype within the TME.

Following the trend in recent years in the use of combination therapies, exploring the synergy between BiTEs and other immunomodulatory drugs may be an ideal strategy to improve efficacy in solid tumors. BiTE therapies in combination with immune checkpoint inhibitors are currently under investigation and may prove to be an effective strategy capable of augmenting immunogenic treatment effects. As we continue to refine our understanding of the tumor immune profile with spatial analyses utilizing genomics and transcriptomics, we will likely be able to strengthen the biologic rationale for combinations therapies and areas of synergy amongst classes of agents.

## 6. Conclusions

Immunotherapy has become an established cornerstone of therapy within the treatment paradigm of several tumor types. Novel T-cell redirecting strategies represent more contemporary immunotherapies, and BiTEs in particular have shown substantial efficacy in hematological malignancies. However, some factors challenge the implementation of BiTEs in solid tumors, namely the lack of target antigen expression, tumor inaccessibility, and the impact of an immunosuppressive tumor microenvironment. Prostate cancer in particular is known for its immune-desert phenotype, with little observed benefit from modern immunotherapies. However, several strategies are currently being investigated to improve BiTE application in an immunologically "cold" tumor and are poised to have transformative impacts within the prostate cancer disease space. Future results of ongoing studies in which the combination of BiTE with innovative therapies may provide some answers and paradigm-changing advances in upcoming years.

**Author Contributions:** D.C.S. and P.C.B. designed the project; D.C.S., J.L.M. and R.L. performed the acquisition, analysis, and interpretation of data; D.C.S. and K.K.Z. wrote the manuscript. J.A.G., W.K.K. and P.C.B. performed the manuscript supervision. All authors have read and agreed to the published version of the manuscript.

**Funding:** This research received no external funding.

**Conflicts of Interest:** The authors declare no conflict of interest related to this work.

## References


1. Kelly, P.N. The Cancer Immunotherapy Revolution. *Science* **2018**, *359*, 1344–1345. [CrossRef] [PubMed]
2. Waldman, A.D.; Fritz, J.M.; Leonardo, M.J. A guide to cancer immunotherapy: From T cell basic science to clinical practice. *Nat. Rev. Immunol.* **2020**, *20*, 651–668. [CrossRef] [PubMed]
3. Oiseth, S.J.; Aziz, M.S. Cancer immunotherapy: A brief review of the history, possibilities, and challenges ahead. *J. Cancer Metastasis Treat* **2017**, *3*, 250–261. [CrossRef]
4. Rosenberg, S.A. IL-2: The first effective immunotherapy for human cancer. *J. Immunol.* **2014**, *192*, 5451–5458. [CrossRef] [PubMed]
5. Skrombolas, D.; Frelinger, J.G. Challenges and developing solutions for increasing the benefits of IL-2 treatment in tumor therapy. *Expert Rev. Clin. Immunol.* **2014**, *10*, 207–217. [CrossRef] [PubMed]
6. Ribas, A.; Wolchok, J.D. Cancer immunotherapy using checkpoint blockade. *Science* **2018**, *359*, 1350–1355. [CrossRef]
7. Robert, C. A decade of immune-checkpoint inhibitors in cancer therapy. *Nat. Commun.* **2020**, *11*, 3801. [CrossRef]
8. Antonarakis, E.S.; Shaukat, F.; Velho, P.I.; Kaur, H.; Shenderov, E.; Pardoll, D.M.; Lotan, T.L. Clinical Features and Therapeutic Outcomes in Men with Advanced Prostate Cancer and DNA Mismatch Repair Gene Mutations. *Eur. Urol.* **2019**, *75*, 378–382. [CrossRef]
9. Kantoff, P.W.; Higano, C.S.; Shore, N.D.; Berger, E.R.; Small, E.J.; Penson, D.F.; Redfern, C.H.; Ferrari, A.C.; Dreicer, R.; Sims, R.B.; et al. Sipuleucel-T immunotherapy for castration-resistant prostate cancer. *N. Engl. J. Med.* **2010**, *363*, 411–422. [CrossRef]
10. Molina, J.C.; Shah, N.S. CAR T cells better than BiTEs. *Blood Adv.* **2021**, *5*, 602–606. [CrossRef]
11. Subklewe, M. BiTEs better than CAR T cells. *Blood Adv.* **2021**, *5*, 607–612. [CrossRef]
12. Sterner, R.C.; Sterner, R.M. CAR-T cell therapy: Current limitations and potential strategies. *Blood Cancer J.* **2021**, *11*, 69. [CrossRef]
13. Fucà, G.; Spagnoletti, A.; Ambrosini, M.; de Braud, F.; Di Nicola, M. Immune cell engagers in solid tumors: Promises and challenges of the next generation immunotherapy. *ESMO Open* **2021**, *6*, 100046. [CrossRef] [PubMed]
14. Nisonoff, A.; Wissler, F.C.; Lipman, L.N. Properties of the major component of a peptic digest of rabbit antibody. *Science* **1960**, *132*, 1770–1771. [CrossRef]
15. Riethmuller, G. Symmetry breaking: Bispecific antibodies, the beginnings, and 50 years on. *Cancer Immun.* **2012**, *12*, 12. [PubMed]
16. Weiner, G.J. Building better monoclonal antibody-based therapeutics. *Nat. Rev. Cancer* **2015**, *15*, 361–370. [CrossRef] [PubMed]
17. Goebeler, M.E.; Bargou, R.C. T cell-engaging therapies—BiTEs and beyond. *Nat. Rev. Clin. Oncol.* **2020**, *17*, 418–434. [CrossRef] [PubMed]
18. Antonarelli, G.; Giugliano, F.; Corti, C.; Repetto, M.; Tarantino, P.; Curigliano, G. Research and Clinical Landscape of Bispecific Antibodies for the Treatment of Solid Malignancies. *Pharmaceuticals* **2021**, *14*, 884. [CrossRef] [PubMed]
19. Brinkmann, U.; Kontermann, R.E. The making of bispecific antibodies. *MAbs* **2017**, *9*, 182–212. [CrossRef]
20. Smith-Garvin, J.E.; Koretzky, G.A.; Jordan, M.S. T cell activation. *Annu. Rev. Immunol.* **2009**, *27*, 591–619. [CrossRef]
21. Huehls, A.M.; Coupet, T.A.; Sentman, C.L. Bispecific T-cell engagers for cancer immunotherapy. *Immunol. Cell Biol.* **2015**, *93*, 290–296. [CrossRef] [PubMed]
22. Einsele, H.; Borghaei, H.; Orlowski, R.Z.; Subklewe, M.; Roboz, G.J.; Zugmaier, G.; Kufer, P.; Iskander, K.; Kantarjian, H.M. The BiTE (bispecific T-cell engager) platform: Development and future potential of a targeted immuno-oncology therapy across tumor types. *Cancer* **2020**, *126*, 3192–3201. [CrossRef] [PubMed]
23. Mack, M.; Riethmüller, G.; Kufer, P. A small bispecific antibody construct expressed as a functional single-chain molecule with high tumor cell cytotoxicity. *Proc. Natl. Acad. Sci. USA* **1995**, *92*, 7021–7025. [CrossRef] [PubMed]
24. Shim, H. Bispecific Antibodies and Antibody-Drug Conjugates for Cancer Therapy: Technological Considerations. *Biomolecules* **2020**, *10*, 360. [CrossRef]
25. Offner, S.; Hofmeister, R.; Romaniuk, A.; Kufer, P.; Baeuerle, P.A. Induction of regular cytolytic T cell synapses by bispecific single-chain antibody constructs on MHC class I-negative tumor cells. *Mol. Immunol.* **2006**, *43*, 763–771. [CrossRef]
26. Bargou, R.; Leo, E.; Zugmaier, G.; Klinger, M.; Goebeler, M.; Knop, S.; Noppeney, R.; Viardot, A.; Hess, G.; Schuler, M.; et al. Tumor regression in cancer patients by very low doses of a T cell-engaging antibody. *Science* **2008**, *321*, 974–977. [CrossRef]
27. Wolf, E.; Hofmeister, R.; Kufer, P.; Schlereth, B.; Baeuerle, P.A. BiTEs: Bispecific antibody constructs with unique anti-tumor activity. *Drug Discov. Today* **2005**, *10*, 1237–1244. [CrossRef]
28. Ross, S.L.; Sherman, M.; McElroy, P.L.; Lofgren, J.A.; Moody, G.; Baeuerle, P.A.; Coxon, A.; Arvedson, T. Bispecific T cell engager (BiTE®) antibody constructs can mediate bystander tumor cell killing. *PLoS ONE* **2017**, *12*, e0183390. [CrossRef]
29. Slaney, C.Y.; Wang, P.; Darcy, P.K.; Kershaw, M.H. CARs versus BiTEs: A Comparison between T Cell-Redirection Strategies for Cancer Treatment. *Cancer Discov.* **2018**, *8*, 924–934. [CrossRef]
30. EMA Blinatumomab. Available online: https://www.ema.europa.eu/en/documents/product-information/blincyto-epar-product-information_en.pdf (accessed on 15 November 2022).
31. FDA Blinatumomab. Available online: https://www.accessdata.fda.gov/drugsatfda_docs/label/2018/125557s013lbl.pdf (accessed on 15 November 2022).
32. Kantarjian, H.; Stein, A.; Gökbuget, N.; Fielding, A.K.; Schuh, A.C.; Ribera, J.M.; Wei, A.; Dombret, H.; Foà, R.; Bassan, R.; et al. Blinatumomab versus Chemotherapy for Advanced Acute Lymphoblastic Leukemia. *N. Engl. J. Med.* **2017**, *376*, 836–847. [CrossRef]

33. Martinelli, G.; Boissel, N.; Chevallier, P.; Ottmann, O.; Gökbuget, N.; Rambaldi, A.; Ritchie, E.K.; Papayannidis, C.; Tuglus, C.A.; Morris, J.D.; et al. Long-term follow-up of blinatumomab in patients with relapsed/refractory Philadelphia chromosome-positive B-cell precursor acute lymphoblastic leukaemia: Final analysis of ALCANTARA study. *Eur. J. Cancer* **2021**, *146*, 107–114. [CrossRef] [PubMed]
34. Gökbuget, N.; Dombret, H.; Bonifacio, M.; Reichle, A.; Graux, C.; Faul, C.; Diedrich, H.; Topp, M.S.; Brüggemann, M.; Horst, H.A.; et al. Blinatumomab for minimal residual disease in adults with B-cell precursor acute lymphoblastic leukemia. *Blood* **2018**, *131*, 1522–1531. [CrossRef] [PubMed]
35. Topp, M.S.; Gökbuget, N.; Stein, A.S.; Zugmaier, G.; O'Brien, S.; Bargou, R.C.; Dombret, H.; Fielding, A.K.; Heffner, L.; Larson, R.A.; et al. Safety and activity of blinatumomab for adult patients with relapsed or refractory B-precursor acute lymphoblastic leukaemia: A multicentre, single-arm, phase 2 study. *Lancet Oncol.* **2015**, *16*, 57–66. [CrossRef] [PubMed]
36. Lee, D.W.; Gardner, R.; Porter, D.L.; Louis, C.U.; Ahmed, N.; Jensen, M.; Grupp, S.A.; Mackall, C.L. Current concepts in the diagnosis and management of cytokine release syndrome. *Blood* **2014**, *124*, 188–195. [CrossRef] [PubMed]
37. Shimabukuro-Vornhagen, A.; Gödel, P.; Subklewe, M.; Stemmler, H.J.; Schlößer, H.A.; Schlaak, M.; Kochanek, M.; Böll, B.; von Bergwelt-Baildon, M.S. Cytokine release syndrome. *J. Immunother. Cancer* **2018**, *6*, 56. [CrossRef]
38. Maus, M.V.; Alexander, S.; Bishop, M.R.; Brudno, J.N.; Callahan, C.; Davila, M.L.; Diamonte, C.; Dietrich, J.; Fitzgerald, J.C.; Frigault, M.J. Society for Immunotherapy of Cancer (SITC) clinical practice guideline on immune effector cell-related adverse events. *J. Immunother. Cancer* **2020**, *8*, e001511. [CrossRef]
39. Morris, E.C.; Neelapu, S.S.; Giavridis, T.; Sadelain, M. Cytokine release syndrome and associated neurotoxicity in cancer immunotherapy. *Nat. Rev. Immunol.* **2022**, *22*, 85–96. [CrossRef]
40. Klinger, M.; Zugmaier, G.; Nägele, V.; Goebeler, M.E.; Brandl, C.; Stelljes, M.; Lassmann, H.; von Stackelberg, A.; Bargou, R.C.; Kufer, P. Adhesion of T Cells to Endothelial Cells Facilitates Blinatumomab-Associated Neurologic Adverse Events. *Cancer Res.* **2020**, *80*, 91–101. [CrossRef]
41. Stein, A.S.; Schiller, G.; Benjamin, R.; Jia, C.; Zhang, A.; Zhu, M.; Zimmerman, Z.; Topp, M.S. Neurologic adverse events in patients with relapsed/refractory acute lymphoblastic leukemia treated with blinatumomab: Management and mitigating factors. *Ann. Hematol.* **2019**, *98*, 159–167. [CrossRef]
42. Li, H.; Er Saw, P.; Song, E. Challenges and strategies for next-generation bispecific antibody-based antitumor therapeutics. *Cell Mol. Immunol.* **2020**, *17*, 451–461. [CrossRef]
43. Lorenczewski, G.; Friedrich, M.; Kischel, R.; Dahlhoff, C.; Anlahr, J.; Balazs, M.; Rock, D.; Boyle, M.; Goldstein, R.; Coxon, A.; et al. Generation of a half-life extended anti-CD19 BiTE® antibody construct compatible with once-weekly dosing for treatment of CD19-positive malignancies. *Blood* **2017**, *130*, 2815. [CrossRef]
44. Sánchez, P.M.; Gordon, P.; Schwartz, S.; Rossi, G.; Huguet, F.; Hernández-Rivas, J.M.; Kadu, P.; Wong, H.L.; Markovic, A.; Katlinskaya, Y.; et al. Safety and Efficacy of Subcutaneous (SC) Blinatumomab for the Treatment of Adults with Relapsed or Refractory B Cell Precursor Acute Lymphoblastic Leukemia (R/R B-ALL). *Blood* **2021**, *138* (Suppl. 1), 2303. [CrossRef]
45. Edeline, J.; Houot, R.; Marabelle, A.; Alcantara, M. CAR-T cells and BiTEs in solid tumors: Challenges and perspectives. *J. Hematol. Oncol.* **2021**, *14*, 65. [CrossRef]
46. Braig, F.; Brandt, A.; Goebeler, M.; Tony, H.P.; Kurze, A.K.; Nollau, P.; Bumm, T.; Böttcher, S.; Bargou, R.C.; Binder, M. Resistance to anti- CD19/CD3 BiTE in acute lymphoblastic leukemia may be mediated by disrupted CD19 membrane trafficking. *Blood* **2017**, *129*, 100–104. [CrossRef] [PubMed]
47. Middelburg, J.; Kemper, K.; Engelberts, P.; Labrijn, A.F.; Schuurman, J.; van Hall, T. Overcoming Challenges for CD3-Bispecific Antibody Therapy in Solid Tumors. *Cancers* **2021**, *13*, 287. [CrossRef]
48. Binnewies, M.; Roberts, E.W.; Kersten, K.; Chan, V.; Fearon, D.F.; Merad, M.; Coussens, L.M.; Gabrilovich, D.I.; Ostrand-Rosenberg, S.; Hedrick, C.C.; et al. Understanding the tumor immune microenvironment (TIME) for effective therapy. *Nat. Med.* **2018**, *24*, 541–550. [CrossRef]
49. Belmontes, B.; Sawant, D.V.; Zhong, W.; Tan, H.; Kaul, A.; Aeffner, F.; O'Brien, S.A.; Chun, M.; Noubade, R.; Eng, J.; et al. Immunotherapy combinations overcome resistance to bispecific T cell engager treatment in T cell-cold solid tumors. *Sci. Transl. Med.* **2021**, *13*, eabd1524. [CrossRef]
50. Arvedson, T.; Bailis, J.M.; Britten, C.D.; Klinger, M.; Nagorsen, N.; Coxon, A.; Egen, J.G.; Martin, F. Targeting Solid Tumors with Bispecific T Cell Engager Immune Therapy. *Annu. Rev. Cancer Biol.* **2022**, *6*, 17–34. [CrossRef]
51. Heiss, M.M.; Murawa, P.; Koralewski, P.; Kutarska, E.; Kolesnik, O.O.; Ivanchenko, V.V.; Dudnichenko, A.S.; Aleknaviciene, B.; Razbadauskas, A.; Gore, M.; et al. The trifunctional antibody catumaxomab for the treatment of malignant ascites due to epithelial cancer: Results of a prospective randomized phase II/III trial. *Int. J. Cancer* **2010**, *127*, 2209–2221. [CrossRef]
52. Chelius, D.; Ruf, P.; Gruber, P.; Plöscher, M.; Liedtke, R.; Gansberger, E.; Hess, J.; Wasiliu, M.; Lindhofer, H. Structural and functional characterization of the trifunctional antibody catumaxomab. *MAbs* **2010**, *2*, 309–319. [CrossRef]
53. Linke, R.; Klein, A.; Seimetz, D. Catumaxomab: Clinical development and future directions. *MAbs* **2010**, *2*, 129–136. [CrossRef]
54. Kebenko, M.; Goebeler, M.E.; Wolf, M.; Hasenburg, A.; Seggewiss-Bernhardt, R.; Ritter, B.; Rautenberg, B.; Atanackovic, D.; Kratzer, A.; Rottman, J.B.; et al. A multicenter phase 1 study of solitomab (MT110, AMG 110), a bispecific EpCAM/CD3 T-cell engager (BiTE®) antibody construct, in patients with refractory solid tumors. *Oncoimmunology* **2018**, *7*, e1450710. [CrossRef] [PubMed]

55. Tabernero, J.; Melero, I.; Ros, W.; Argiles, G.; Marabelle, A.; Rodriguez-Ruiz, M.E.; Albanell, J.; Calvo, E.; Moreno, V.; Cleary, J.M. Phase Ia and Ib studies of the novel carcinoembryonic antigen (CEA) T-cell bispecific (CEA CD3 TCB) antibody as a single agent and in combination with atezolizumab: Preliminary efficacy and safety in patients with metastatic colorectal cancer (mCRC). *J. Clin. Oncol.* **2017**, *35*, 3002. [CrossRef]
56. El-Rayes, B.; Hendifar, A.E.; Pant, S.; Wilky, B.A.; Reilley, M.; Benson, A.B.; Chow, W.A.; Konda, B.; Starr, J.; Ahn, D.H.; et al. Preliminary Safety, Pharmacodynamic, and Antitumor Activity of Tidutamab, an SSTR2 x CD3 Bispecific Antibody, in Subjects with Advanced Neuroendocrine Tumors. In Proceedings of the 2021 NANETS Annual Symposium, Virtual, 3–6 November 2021. Abstract 109.
57. Pietzner, K.; Vergote, I.; Santoro, A.; Marme, F.; Rosenberg, P.; Friccius-Quecke, H.; Sehouli, J. Results of a phase II clinical trial to evaluate a re-challenge of intraperitoneal catumaxomab for treatment of malignant ascites (MA) due to epithelial cancer (SECIMAS). *J. Clin. Oncol.* **2013**, *31*, 5582. [CrossRef]
58. Berek, J.S.; Edwards, R.P.; Parker, L.P.; DeMars, L.R.; Herzog, T.J.; Lentz, S.S.; Morris, R.T.; Akerley, W.L.; Holloway, R.W.; Method, M.W.; et al. Catumaxomab for the treatment of malignant ascites in patients with chemotherapy-refractory ovarian cancer: A phase II study. *Int. J. Gynecol. Cancer* **2014**, *24*, 1583–1589. [CrossRef] [PubMed]
59. Romero, I.; Oaknin, A.; Arranz, J.A.; García-Martínez, E.; Herrero, A.; Casado, A.; De Juan, A.; Guerra, E.; Polo, S.H.; Santaballa, A.; et al. Phase II trial of intraperitoneal (IP) administration of catumaxomab (C) as consolidation therapy for patients (pts) with relapsed epithelial ovarian cancer (OC) in second or third complete remission: GEICO 1001 study. *J. Clin. Oncol.* **2014**, *32*, 5528. [CrossRef]
60. Baumann, K.; Pfisterer, J.; Wimberger, P.; Burchardi, N.; Kurzeder, C.; du Bois, A.; Loibl, S.; Sehouli, J.; Huober, J.; Schmalfeldt, B.; et al. Intraperitoneal treatment with the trifunctional bispecific antibody Catumaxomab in patients with platinum-resistant epithelial ovarian cancer: A phase IIa study of the AGO Study Group. *Gynecol. Oncol.* **2011**, *123*, 27–32. [CrossRef]
61. Lordick, F.; Kunzmann, V.; Trojan, J.; Daum, S.; Schenk, M.; Kullmann, F.; Schroll, S.; Behringer, D.M.; Stahl, M.; Al-Batran, S.-E.; et al. Intraperitoneal immunotherapy with the bispecific anti-EpCAM x anti-CD3 directed antibody catumaxomab for patients with peritoneal carcinomatosis from gastric cancer: Final results of a randomized phase II AIO trial. *J. Clin. Oncol.* **2018**, *4*, 4. [CrossRef]
62. Pishvaian, M.J.; Morse, M.; McDevitt, J.T.; Ren, S.; Robbie, G.; Ryan, P.C.; Soukharev, S.; Bao, H.; Denlinger, C.S. Phase 1 dose escalation study of MEDI-565, a bispecific T-cell engager that targets human carcinoembryonic antigen (CEA), in patients with advanced gastrointestinal (GI) adenocarcinomas. *J. Clin. Oncol.* **2016**, *34*, 320. [CrossRef]
63. Moek, K.L.; Fiedler, W.M.; von Einem, J.C.; Verheul, H.M.; Seufferlein, T.; de Groot, D.J.; Heinemann, V.; Kebenko, M.; Menke-van der Houven van Oordt, C.M.; Ettrich, T.J.; et al. Phase I study of AMG 211/MEDI-565 administered as continuous intravenous infusion (cIV) for relapsed/refractory gastrointestinal (GI) adenocarcinoma. *Ann. Oncol.* **2018**, *29*, VIII139–VIII140. [CrossRef]
64. Johnson, M.L.; Solomon, B.J.; Awad, M.M.; Cho, B.C.; Gainor, J.F.; Goldberg, S.B.; Keam, B.; Lee, D.H.; Huang, C.; Helms, H.-J.; et al. MORPHEUS: A phase Ib/II multi-trial platform evaluating the safety and efficacy of cancer immunotherapy (CIT)-based combinations in patients (pts) with non-small cell lung cancer (NSCLC). *J. Clin. Oncol.* **2018**, *36*, TPS9105. [CrossRef]
65. Lum, L.G.; Thakur, A.; Choi, M.; Deol, A.; Kondadasula, V.; Schalk, D.; Fields, K.; Dufrense, M.; Philip, P.; Dyson, G.; et al. Clinical and immune responses to anti-CD3 x anti-EGFR bispecific antibody armed activated T cells (EGFR BATs) in pancreatic cancer patients. *Oncoimmunology* **2020**, *9*, 1773201. [CrossRef] [PubMed]
66. Rosenthal, M.; Balana, C.; Van Linde, M.E.; Sayehli, C.; Fiedler, W.M.; Wermke, M.; Massard, C.; Ang, A.; Kast, J.; Stienen, S.; et al. Novel anti-EGFRvIII bispecific T cell engager (BiTE) antibody construct in glioblastoma (GBM): Trial in progress of AMG 596 in patients with recurrent or newly diagnosed disease. *J. Clin. Oncol.* **2019**, *37*, TPS2071. [CrossRef]
67. Powderly, J.D.; Hurwitz, H.; Ryan, D.P.; Laheru, D.A.; Pandya, N.B.; Lohr, J.; Moore, P.A.; Bonvini, E.; Wigginton, J.M.; Crocenzi, T.S. A phase 1, first-in-human, open label, dose escalation study of MGD007, a humanized gpA33 x CD3 DART molecule, in patients with relapsed/refractory metastatic colorectal carcinoma. *J. Clin. Oncol.* **2016**, *34*, TPS3628. [CrossRef]
68. Owonikoko, T.K.; Champiat, S.; Johnson, M.L.; Govindan, R.; Hiroki Izumi, H.; Lai, V.W.; Borghaei, H.; Boyer, M.J.; Boosman, R.J.; Hummel, H.-D.; et al. Updated results from a phase 1 study of AMG 757, a half-life extended bispecific T-cell engager (BiTE) immuno-oncology therapy against delta-like ligand 3 (DLL3), in small cell lung cancer (SCLC). *J. Clin. Oncol.* **2021**, *39*, 8510. [CrossRef]
69. Winer, I.S.; Shields, A.F.; Yeku, O.O.; Liu, J.F.; Peterman, M.J.; Yoo, S.Y.; Lowy, I.; Yama-Dang, N.A.; Goncalves, P.H.; Kroog, G. A phase I/II, multicenter, open-label study of REGN5668 (mucin [MUC]16 x CD28 bispecific antibody [bsAb]) with cemiplimab (programmed death [PD]-1 Ab) or REGN4018 (MUC16 x CD3 bsAb) in recurrent ovarian cancer (rOVCA). *J. Clin. Oncol.* **2021**, *39*, TPS5602. [CrossRef]
70. EMA Tebentafusp. Available online: https://www.ema.europa.eu/en/documents/assessment-report/kimmtrak-epar-public-assessment-report_en.pdf (accessed on 26 November 2022).
71. FDA Tebentafusp. Available online: https://www.accessdata.fda.gov/drugsatfda_docs/label/2022/761228s000lbl.pdf (accessed on 26 November 2022).
72. Nathan, P.; Hassel, J.C.; Rutkowski, P.; Baurain, J.F.; Butler, M.O.; Schlaak, M.; Sullivan, R.J.; Ochsenreither, S.; Dummer, R.; Kirkwood, J.M.; et al. Overall Survival Benefit with Tebentafusp in Metastatic Uveal Melanoma. *N. Engl. J. Med.* **2021**, *385*, 1196–1206. [CrossRef]

73. Madan, R.A.; Antonarakis, E.S.; Drake, C.G.; Fong, L.; Yu, E.Y.; McNeel, D.G.; Lin, D.W.; Chang, N.N.; Sheikh, N.A.; Gulley, J.L. Putting the Pieces Together: Completing the Mechanism of Action Jigsaw for Sipuleucel-T. *J. Natl. Cancer Inst.* **2020**, *112*, 562–573. [CrossRef]
74. Bonaventura, P.; Shekarian, T.; Alcazer, V.; Valladeau-Guilemond, J.; Valsesia-Wittmann, S.; Amigorena, S.; Caux, C.; Depil, S. Cold Tumors: A Therapeutic Challenge for Immunotherapy. *Front. Immunol.* **2019**, *10*, 168. [CrossRef]
75. Mehra, N.; Seed, G.; Lambros, M.; Sharp, A.; Fontes, M.S.; Crespo, M.; Sumanasuriva, S.; Yuan, W.; Bovsen, G.; Riisnaes, R.; et al. Myeloid-derived suppressor cells (MDSCs) in metastatic castration-resistant prostate cancer (CRPC) patients (PTS). *Ann. Oncol.* **2016**, *27*, vi257. [CrossRef]
76. Kfoury, Y.; Baryawno, N.; Severe, N.; Mei, S.; Gustafsson, K.; Hirz, T.; Brouse, T.; Scadden, E.W.; Igolkina, A.A.; Kokkaliaris, K.; et al. Human prostate cancer bone metastases have an actionable immunosuppressive microenvironment. *Cancer Cell* **2021**, *39*, 1464–1478. [CrossRef]
77. Guan, X.; Polesso, F.; Wang, C.; Sehrawat, A.; Hawkins, R.M.; Murray, S.E.; Thomas, G.V.; Caruso, B.; Thompson, R.F.; Wood, M.A.; et al. Androgen receptor activity in T cells limits checkpoint blockade efficacy. *Nature* **2022**, *606*, 791–796. [CrossRef] [PubMed]
78. Dorff, T.B.; Narayan, V.; Forman, S.J.; Zang, P.D.; Fraietta, J.A.; June, C.H.; Haas, N.B.; Priceman, S.J. Novel Redirected T–Cell Immunotherapies for Advanced Prostate Cancer. *Clin. Cancer Res.* **2022**, *28*, 576–584. [CrossRef]
79. Hummel, H.D.; Kufer, P.; Grüllich, C.; Seggewiss-Bernhardt, R.; Deschler-Baier, B.; Chatterjee, M.; Goebeler, M.E.; Miller, K.; de Santis, M.; Loidl, W.; et al. Pasotuxizumab, a BiTE® immune therapy for castration-resistant prostate cancer: Phase I, dose-escalation study findings. *Immunotherapy* **2021**, *13*, 125–141. [CrossRef] [PubMed]
80. Hummel, H.D.; Kufer, P.; Grullich, C.; Deschler-Baier, B.; Chatterjee, M.; Goebeler, M.E.; Miller, K.; De Santis, M.; Loidl, W.C.; Buck, A.; et al. Phase 1 study of pasotuxizumab (BAY 2010112), a PSMA-targeting Bispecific T cell Engager (BiTE) immunotherapy for metastatic castration resistant prostate cancer (mCRPC). *J. Clin. Oncol.* **2019**, *37*, 5034. [CrossRef]
81. Paweletz, K.L.; Li, S.; Bailis, J.M.; Juan, G. Combination of AMG 160, a PSMA x CD3 half-life extended bispecific T-cell engager (HLE BiTE) immune therapy, with an anti-PD-1 antibody in prostate cancer (PCa). *J. Clin. Oncol.* **2020**, *38*, 155. [CrossRef]
82. Tran, B.; Horvath, L.; Dorff, T.; Retting, T.; Lolkema, M.P.; Machiels, J.; Rottey, S.; Autio, K.; Greil, R.; Adra, N.; et al. 6090-Results from a phase I study of AMG 160, a half-life extended (HLE), PSMA-targeted, bispecific T-cell engager (BiTE) immune therapy for metastatic castration-resistant prostate cancer (mCRPC). *Ann. Oncol.* **2020**, *31*, S507–S549. [CrossRef]
83. Subudhi, S.K.; Siddiqui, B.A.; Maly, J.J.; Nandagopal, L.; Lam, E.T.; Whang, Y.E.; Minocha, M.; Gupta, V.; Penny, X.; Cooner, F.; et al. Safety and efficacy of AMG 160, a half-life extended BiTE immune therapy targeting prostate-specific membrane antigen (PSMA), and other therapies for metastatic castration-resistant prostate cancer (mCRPC). *J. Clin. Oncol.* **2021**, *39*, TPS5088. [CrossRef]
84. Hernandez-Hoyos, G.; Sewell, T.; Bader, R.; Bannink, J.; Chenault, R.A.; Daugherty, M.; Dasovich, M.; Fang, H.; Gottschalk, R.; Kumer, J.; et al. MOR209/ES414, a Novel Bispecific Antibody Targeting PSMA for the Treatment of Metastatic Castration-Resistant Prostate Cancer. *Mol. Cancer* **2016**, *15*, 2155–2165. [CrossRef]
85. De Bono, J.S.; Fong, L.; Beer, T.M.; Gao, X.; Geynisman, D.M.; Burris, H.A.; Strauss, J.F.; Courtney, K.D.; Quinn, D.I.; VanderWeele, D.J.; et al. Results of an ongoing phase 1/2a dose escalation study of HPN424, a tri-specific half-life extended PSMA-targeting T-cell engager, in patients with metastatic castration-resistant prostate cancer (mCRPC). *J. Clin. Oncol.* **2021**, *39*, 5013. [CrossRef]
86. Lim, E.A.; Schweizer, M.T.; Chi, K.N.; Aggarwal, R.R.; Agarwal, N.; Gulley, J.L.; Attiyeh, E.F.; Greger, J.; Wu, S.; Jaiprasart, P.; et al. Safety and preliminary clinical activity of JNJ-63898081 (JNJ-081), a PSMA and CD3 bispecific antibody, for the treatment of metastatic castrate-resistant prostate cancer (mCRPC). *J. Clin. Oncol.* **2022**, *40*, 279. [CrossRef]
87. Heitmann, J.S.; Walz, J.S.; Pflügler, M.; Marconato, M.; Tegeler, C.M.; Reusch, J.; Labrenz, J.; Schlenk, R.; Jung, G.; Salih, H. Abstract CT141: CC-1, a bispecific PSMAxCD3 antibody for treatment of prostate carcinoma: Results of the ongoing phase I dose escalation trial. *Cancer Res.* **2022**, *82*, CT141. [CrossRef]
88. Markowski, M.C.; Kilari, D.; Eisenberger, M.A.; McKay, R.R.; Dreicer, R.; Trikha, M.; Heath, E.I.; Li, J.; Garzone, P.D.; Young, T.S. Phase I study of CCW702, a bispecific small molecule-antibody conjugate targeting PSMA and CD3 in patients with metastatic castration-resistant prostate cancer (mCRPC). *J. Clin. Oncol.* **2021**, *39*, TPS5094. [CrossRef]
89. Kelly, W.K.; Thanigaimani, P.; Sun, F.; Seebach, F.A.; Lowy, I.; Sandigursky, S.; Miller, E. A phase 1/2 study of REGN4336, a PSMAxCD3 bispecific antibody, alone and in combination with cemiplimab in patients with metastatic castration-resistant prostate cancer. *J. Clin. Oncol.* **2022**, *40*, TPS5105. [CrossRef]
90. Buelow, B.; Dalvi, P.; Dang, K.; Patel, A.; Johal, K.; Pham, D.; Panchal, S.; Liu, Y.; Fong, L.; Sartor, A.O.; et al. TNB585.001: A multicenter, phase 1, open-label, dose-escalation and expansion study of tnb-585, a bispecific T-cell engager targeting PSMA in subjects with metastatic castrate resistant prostate cancer. *J. Clin. Oncol.* **2021**, *39*, TPS5092. [CrossRef]
91. Danila, D.C.; Waterhouse, D.M.; Appleman, L.J.; Pook, D.W.; Matsubara, N.; Dorff, T.B.; Lee, J.; Armstrong, A.J.; Kim, M.; Horvath, L.; et al. A phase 1 study of AMG 509 in patients (pts) with metastatic castration-resistant prostate cancer (mCRPC). *J. Clin. Oncol.* **2022**, *40*, TPS5101. [CrossRef]
92. Aggarwal, R.R.; Aparicio, A.; Heidenreich, A.; Sandhu, S.K.; Zhang, Y.; Salvati, M.; Shetty, A.; Sadraei, N.H. Phase 1b study of AMG 757, a half-life extended bispecific T-cell engager (HLE BiTEimmune-oncology therapy) targeting DLL3, in de novo or treatment emergent neuroendocrine prostate cancer (NEPC). *J. Clin. Oncol.* **2021**, *39*, TPS5100. [CrossRef]

93. Patel, M.; Lum, L.G.; Deol, A.; Thakur, A.; Heath, E.I.; Chen, W.; Dobson, K.; Fontana, J.A.; Vaishampayan, U.N. Phase II trial of a novel immunotherapy combination of pembrolizumab and HER2 bi-armed activated T cells (BATs) in metastatic castrate resistant prostate cancer. *J. Clin. Oncol.* **2020**, *38*, 97. [CrossRef]
94. Silver, D.A.; Pellicer, I.; Fair, W.R.; Heston, W.D.; Cordon-Cardo, C. Prostate-specific membrane antigen expression in normal and malignant human tissues. *Clin. Cancer Res.* **1997**, *3*, 81–85.
95. Mhawech-Fauceglia, P.; Zhang, S.; Terracciano, L.; Sauter, G.; Chadhuri, A.; Herrmann, F.R.; Penetrante, R. Prostate-specific membrane antigen (PSMA) protein expression in normal and neoplastic tissues and its sensitivity and specificity in prostate adenocarcinoma: An immunohistochemical study using mutiple tumour tissue microarray technique. *Histopathology* **2007**, *50*, 472–483. [CrossRef]
96. Chang, S.S.; Reuter, V.E.; Heston, W.D.; Bander, N.H.; Grauer, L.S.; Gaudin, P.B. Five different anti-prostate-specific membrane antigen (PSMA) antibodies confirm PSMA expression in tumor-associated neovasculature. *Cancer Res.* **1999**, *59*, 3192–3198. [PubMed]
97. Van de Wiele, C.; Sathekge, M.; de Spiegeleer, B.; De Jonghe, P.J.; Debruyne, P.R.; Borms, M.; Beels, L.; Maes, A. PSMA expression on neovasculature of solid tumors. *Histol. Histopathol.* **2020**, *35*, 919–927. [CrossRef]
98. Perera, M.; Papa, N.; Christidis, D.; Wetherell, D.; Hofman, M.S.; Murphy, D.G.; Bolton, D.; Lawrentschuk, N. Sensitivity, Specificity, and Predictors of Positive (68) Ga-Prostate-specific Membrane Antigen Positron Emission Tomography in Advanced Prostate Cancer: A Systematic Review and Meta-analysis. *Eur. Urol.* **2016**, *70*, 926–937. [CrossRef] [PubMed]
99. Morris, M.J.; Rowe, S.P.; Gorin, M.A.; Saperstein, L.; Pouliot, F.; Josephson, D.; Wong, J.Y.C.; Pantel, A.R.; Cho, S.Y.; Gage, K.L.; et al. Diagnostic Performance of $^{18}$F-DCFPyL-PET/CT in Men with Biochemically Recurrent Prostate Cancer: Results from the CONDOR Phase III, Multicenter Study. *Clin. Cancer Res.* **2021**, *27*, 3674–3682. [CrossRef] [PubMed]
100. Sartor, O.; de Bono, J.; Chi, K.N.; Fizazi, K.; Herrmann, K.; Rahbar, K.; Tagawa, S.T.; Nordquist, L.T.; Vaishampayan, N.; El-Haddad, G.; et al. Lutetium-177-PSMA-617 for Metastatic Castration-Resistant Prostate Cancer. *N. Engl. J. Med.* **2021**, *385*, 1091–1103. [CrossRef]
101. Buteau, J.P.; Martin, A.J.; Emmett, L.; Iravani, A.; Sandhu, S.; Joshua, A.M.; Francis, R.J.; Zhang, A.Y.; Scott, A.M.; Lee, S.T.; et al. PSMA and FDG-PET as predictive and prognostic biomarkers in patients given [$^{177}$Lu]Lu-PSMA-617 versus cabazitaxel for metastatic castration-resistant prostate cancer (TheraP): A biomarker analysis from a randomised, open-label, phase 2 trial. *Lancet Oncol.* **2022**, *23*, 1389–1397. [CrossRef]
102. Bailis, J.; Deegen, P.; Thomas, O.; Bogner, P.; Wahl, J.; Liao, M.; Li, S.; Matthes, K.; Nägele, V.; Rau, D.; et al. Preclinical evaluation of AMG 160, a next-generation bispecific T cell engager (BiTE) targeting the prostate-specific membrane antigen PSMA for metastatic castration-resistant prostate cancer (mCRPC). *J. Clin. Oncol.* **2019**, *37*, 301. [CrossRef]
103. Tran, B.; Horvath, L.; Dorff, T.B.; Greil, R.; Machiels, J.-P.H.; Roncolato, F.; Scott, T.; Tagawa, S.T.; Shariat, S.F.; Salvati, M.; et al. Phase I study of AMG 160, a half-life extended bispecific T-cell engager (HLE BiTE) immune therapy targeting prostatespecific membrane antigen (PSMA), in patients with metastatic castration-resistant prostate cancer (mCRPC). *J. Clin. Oncol.* **2020**, *38*, TPS261. [CrossRef]
104. Lemon, B.; Aaron, W.; Austin, R.; Baeuerle, P.; Barath, M.; Jones, A.; Jones, S.D.; Kwant, K.; Law, C.L.; Muchnik, A.; et al. Abstract 1773: HPN424, a half-life extended, PSMA/CD3-specific TriTAC for the treatment of metastatic prostate cancer. *Cancer Res.* **2018**, *78*, 1773. [CrossRef]
105. Raff, A.B.; Gray, A.; Kast, W.M. Prostate stem cell antigen: A prospective therapeutic and diagnostic target. *Cancer Lett.* **2009**, *277*, 126–132. [CrossRef] [PubMed]
106. Saeki, N.; Gu, J.; Yoshida, T.; Wu, X. Prostate stem cell antigen: A Jekyll and Hyde molecule? *Clin. Cancer Res.* **2010**, *16*, 3533–3538. [CrossRef] [PubMed]
107. Gu, Z.; Thomas, G.; Yamashiro, J.; Shintaku, I.P.; Dorey, F.; Raitano, A.; Witte, O.N.; Said, J.W.; Loda, M.; Reiter, R.E. Prostate stem cell antigen (PSCA) expression increases with high gleason score, advanced stage and bone metastasis in prostate cancer. *Oncogene* **2000**, *19*, 1288–1296. [CrossRef] [PubMed]
108. Burnell, S.; Spencer-Harty, S.; Howarth, S.; Bodger, O.; Kynaston, H.; Morgan, C.; Doak, S.H. Utilisation of the STEAP protein family in a diagnostic setting may provide a more comprehensive prognosis of prostate cancer. *PLoS ONE* **2019**, *14*, e0220456. [CrossRef] [PubMed]
109. Zaffuto, E.; Pompe, R.; Zanaty, M.; Bondarenko, H.D.; Leyh-Bannurah, S.R.; Moschini, M.; Dell'Oglio, P.; Gandaglia, G.; Fossati, N.; Stabile, A.; et al. Contemporary Incidence and Cancer Control Outcomes of Primary Neuroendocrine Prostate Cancer: A SEER Database Analysis. *Clin. Genitourin. Cancer* **2017**, *15*, e793–e800. [CrossRef]
110. Bluemn, E.G.; Coleman, I.M.; Lucas, J.M.; Coleman, R.T.; Hernandez-Lopez, S.; Tharakan, R.; Bianchi-Frias, D.; Dumpit, R.F.; Kaipainen, A.; Corella, A.N.; et al. Androgen Receptor Pathway-Independent Prostate Cancer Is Sustained through FGF Signaling. *Cancer Cell* **2017**, *32*, 474–489.e6. [CrossRef] [PubMed]
111. Netto, G.J.; Amin, M.B.; Kench, J.G.; Kench, J.G.; Amin, M.B.; Berney, D.M.; Compérat, E.M.; Cree, I.A.; Grill, A.J.; Hartmann, A.; et al. Chapter 4: Tumours of the Prostate. In *WHO Classification of Tumours*, 5th ed.; Srigley, J.R., Amin, M.B., Rubin, M.A., Tsuzuki, T., Eds.; International Agency for Research on Cancer: Lyon, France, 2022; Volume 8, ISBN 978-92-832-4512-4. Available online: https://tumourclassification.iarc.who.int/chapters/36 (accessed on 3 December 2022).

112. Kench, J.G.; Amin, M.B.; Berney, D.M.; Compérat, E.M.; Cree, I.A.; Gill, A.J.; Hartmann, A.; Menon, S.; Moch, H.; Netto, G.J.; et al. WHO Classification of Tumours fifth edition: Evolving issues in the classification, diagnosis, and prognostication of prostate cancer. *Histopathology* **2022**, *81*, 447–458. [CrossRef]
113. Wang, Y.; Wang, Y.; Ci, X.; Choi, S.Y.C.; Crea, F.; Lin, D.; Wang, Y. Molecular events in neuroendocrine prostate cancer development. *Nat. Rev. Urol.* **2021**, *18*, 581–596. [CrossRef]
114. Puca, L.; Gavyert, K.; Sailer, V.; Conteduca, V.; Dardenne, E.; Sigouros, M.; Isse, K.; Kearney, M.; Vosoughi, A.; Fernandez, L.; et al. Delta-like protein 3 expression and therapeutic targeting in neuroendocrine prostate cancer. *Sci. Transl. Med.* **2019**, *11*, eaav0891. [CrossRef]
115. Han, M.; Li, F.; Zhang, Y.; Dai, P.; He, J.; Li, Y.; Zhu, Y.; Zheng, J.; Huang, H.; Bai, F.; et al. FOXA2 drives lineage plasticity and KIT pathway activation in neuroendocrine prostate cancer. *Cancer Cell* **2022**, *40*, 1306–1323.e8. [CrossRef]
116. Kauer, J.; Hörner, S.; Osburg, L.; Müller, S.; Märklin, M.; Heitmann, J.S.; Zekri, L.; Rammensee, H.G.; Salih, H.R.; Jung, G. Tocilizumab, but not dexamethasone, prevents CRS without affecting antitumor activity of bispecific antibodies. *J. Immunother. Cancer* **2020**, *8*, e000621. [CrossRef]
117. Kamat, N.V.; Yu, E.Y.; Lee, J.K. BiTE-ing into Prostate Cancer with Bispecific T-cell Engagers. *Clin. Cancer Res.* **2021**, *27*, 2675–2677. [CrossRef] [PubMed]

cancers

*Article*

# Subsequent Upper Urinary Tract Carcinoma Related to Worse Survival in Patients Treated with BCG

**Kazuyuki Numakura [1,*], Makito Miyake [2], Mizuki Kobayashi [1], Yumina Muto [1], Yuya Sekine [1], Nobutaka Nishimura [2], Kota Iida [2], Masanori Shiga [3], Shuichi Morizane [4], Takahiro Yoneyama [5], Yoshiaki Matsumura [6], Takashige Abe [7], Takeshi Yamada [8], Kazumasa Matsumoto [9], Junichi Inokuchi [10], Naotaka Nishiyama [11], Rikiya Taoka [12], Takashi Kobayashi [13], Takahiro Kojima [14], Hiroshi Kitamura [11], Hiroyuki Nishiyama [3], Kiyohide Fujimoto [2] and Tomonori Habuchi [1]**

1 Department of Urology, Akita University Graduate School of Medicine, Akita 010-8543, Japan
2 Department of Urology, Nara Medical University, Kashihara 634-8521, Japan
3 Department of Urology, Faculty of Medicine, University of Tsukuba, Tsukuba 305-8576, Japan
4 Division of Urology, Faculty of Medicine, Tottori University, Tottori 683-8504, Japan
5 Department of Urology, Hirosaki University Graduate School of Medicine, Hirosaki 036-8563, Japan
6 Department of Urology, Nara Prefecture General Medical Center, Nara 630-8054, Japan
7 Department of Renal and Genitourinary Surgery, Graduate School of Medicine, Hokkaido University, Sapporo 060-8648, Japan
8 Department of Urology, Kyoto Prefectural University of Medicine, Kyoto 602-8566, Japan
9 Department of Urology, Kitasato University School of Medicine, Sagamihara 252-0373, Japan
10 Department of Urology, Graduate School of Medical Sciences, Kyushu University, Fukuoka 812-8582, Japan
11 Department of Urology, Faculty of Medicine, University of Toyama, Toyama 930-0194, Japan
12 Departments of Urology, Kagawa University Faculty of Medicine, Takamatsu 761-0793, Japan
13 Department of Urology, Kyoto University Graduate School of Medicine, Kyoto 606-8507, Japan
14 Department of Urology, Aichi Cancer Center, Nagoya 464-8681, Japan
* Correspondence: nqf38647@nifty.com; Tel.: +81-18-884-6460

**Simple Summary:** This retrospective cohort study aimed to understand the incidence, clinical impact, and risk factors associated with upper urinary tract urothelial carcinoma (UTUC) after intravesical Bacillus Calmette-Guerin (BCG) therapy. The study included 3226 patients diagnosed with non-muscle-invasive bladder cancer (NMIBC) and treated with intravesical BCG therapy between January 2000 and December 2019. Of these patients, 6.1% were diagnosed with UTUC during the follow-up period, and those with UTUC had worse survival rates compared to those without UTUC. Tumor multiplicity, treatment for Connaught strain, and intravesical recurrence after BCG therapy were associated with subsequent UTUC diagnosis. The study suggests that patients with these risk factors may require closer monitoring for UTUC after BCG therapy.

**Abstract:** Upper urinary tract urothelial carcinoma (UTUC) after intravesical bacillus Calmette-Guerin (BCG) therapy is rare, and its incidence, clinical impact, and risk factors are not fully understood. To elucidate the clinical implications of UTUC after intravesical BCG therapy, this retrospective cohort study used data collected between January 2000 and December 2019. A total of 3226 patients diagnosed with non-muscle-invasive bladder cancer (NMIBC) and treated with intravesical BCG therapy were enrolled (JUOG-UC 1901). UTUC impact was evaluated by comparing intravesical recurrence-free survival (RFS), cancer-specific survival (CSS), and overall survival (OS) rates. The predictors of UTUC after BCG treatment were assessed. Of these patients, 2873 with a medical history that checked UTUC were analyzed. UTUC was detected in 175 patients (6.1%) during the follow-up period. Patients with UTUC had worse survival rates than those without UTUC. Multivariate analyses revealed that tumor multiplicity (odds ratio [OR], 1.681; 95% confidence interval [CI], 1.005–2.812; $p = 0.048$), Connaught strain (OR, 2.211; 95% CI, 1.380–3.543; $p = 0.001$), and intravesical recurrence (OR, 5.097; 95% CI, 3.225–8.056; $p < 0.001$) were associated with UTUC after BCG therapy. In conclusion, patients with subsequent UTUC had worse RFS, CSS, and OS than those without UTUC. Multiple bladder tumors, treatment for Connaught strain, and intravesical recurrence after BCG therapy may be predictive factors for subsequent UTUC diagnosis.

**Keywords:** BCG; non-muscle-invasive bladder cancer; upper urinary tract cancer; intravesical recurrence; multiple bladder tumors

## 1. Introduction

The urinary bladder is the predominant organ affected in primary urothelial carcinoma (UC), although UC can arise throughout the urinary tract. Upper urinary tract urothelial carcinoma (UTUC) accounts for 5% to 10% of all urothelial carcinomas [1]. Although UTUC and UC in the bladder (UCB) have previously been considered to share histopathological features, remarkable clinical and molecular differences exist between cancer sites [2]. Approximately 60% of patients with UTUC have an invasive tumor versus 20% to 25% of patients with UCB [3]. The prognosis of UTUC is poor, with a 5-year overall survival (OS) of approximately 70%. For invasive disease, the 5-year OS is less than 40%, which is worse than the 5-year OS in patients with muscle-invasive UBC treated with radical cystectomy [4,5].

After intravesical bacillus Calmette-Guerin (BCG) instillation therapy, UTUC is diagnosed in 4%–9% of patients [3,6]. This rate was higher than that of UTUC among all patients with UBC [7] and was equivalent to the recurrence rate after total cystectomy [8,9]. Clinical risk factors for the development of subsequent UTUC are tumor multiplicity, tumor location (involved in the ureteral orifice), advanced tumor stage, the existence of carcinoma in situ (CIS), and operative modality [10]. Although every guideline (the National Comprehensive Cancer Network, the American Urological Association, and the European Association of Urology) recommends careful follow-up of the entire urinary tract after treatment for high-grade UC, diagnosis of subsequent UTUC has often been delayed [11]. Indeed, little is known about the clinical implications of subsequent UTUC after BCG instillation for UBC.

In this study, we attempted to elucidate the clinical features of subsequent UTUC after BCG therapy and to elucidate the risk factors for its development in a retrospective cohort of 3226 patients in Japan (JUOG-UC 1901).

## 2. Materials and Methods

### 2.1. Data Collection and Study Cohort

This retrospective multicenter study was approved by the institutional review board of each participating institute (reference protocol ID: 2266 in the IRB of Akita University) of the Japan Urological Oncology Group framework. Informed consent was obtained from the participants through posters and/or websites following the opt-out policy (https://www.lifescience.mext.go.jp/bioethics/seimeikagaku_igaku.html) (last access date 19 March 2023). We recruited 3226 patients who received intravesical BCG therapy for NMIBC treated between 2000 and 2019 at 31 hospitals in Japan. The clinical characteristics of the patients were investigated, including age; sex; performance status; former history of NMIBC; tumor multiplicity; tumor size; T category; tumor grade (per the 2004 World Health Organization classification); second transurethral resection (TUR); presence of bladder CIS and prostatic urethra-involving CIS; divergent differentiation such as squamous differentiation and glandular differentiation; variant histology such as nested, micropapillary, and plasmacytoid variants; and lymphovascular involvement (LVI).

Of the 3226 patients, 353 (11%) were excluded due to missing data on subsequent UTUC status; thus, 2873 (89%) were analyzed. Supplemental Figure S1 shows the flowchart of the patient selection process.

### 2.2. Intravesical BCG Treatment after Transurethral Resection of Bladder Tumor (TURBT)

The criteria, dose, and schedule for initial BCG and maintenance BCG therapy were not strictly scheduled and were given as clinical implementation by each physician's decision. Most eligible patients were at high or highest risk of NMIBC, such as those who had papillary Ta/T1 high-grade tumors and CIS, and were treated with intravesical BCG therapy after TURBT.

The intravesical BCG therapy consisted of weekly instillations of Immunobladder (80 mg of Tokyo-172 strain) or ImmuCyst (81 mg of Connaught strain, currently unavailable) for 6–8 consecutive weeks, with or without subsequent maintenance BCG therapy.

### 2.3. Surveillance

The surveillance protocol varied depending on the policies of the individual institutions and the number of physicians. In general, patients underwent check-ups by white-light cystoscopy and urinary cytology every 3 months for the first 2 years, every 4 months in the third year, every 6 months in the fourth and fifth years, and annually thereafter. This follow-up protocol followed a conventional Japanese style and may be stricter than the established guidelines [12–14]. Recurrence was defined as recurrent tumors of pathologically proven urothelial carcinoma in the bladder and/or prostatic urethra. Progression was defined as recurrent disease with invasion of the muscularis propria (≥T2), positive regional lymph nodes, and/or distant metastases. UTUC was diagnosed based on a positive finding on a computed tomography scan of the upper urinary tract (UUT), ipsilateral positive urine cytology, and ureteroscopic pathological diagnosis.

### 2.4. Statistical Analysis

Clinicopathological characteristics were compared using Mann–Whitney U, chi-square, and Kruskal–Wallis tests as appropriate. Intravesical recurrence-free survival (RFS), cancer-specific survival (CSS), and OS were calculated from the date of initial induction of BCG treatment. Survival rates were analyzed using the Kaplan–Meier method, and significance was compared using the Cox hazard regression model. To elucidate the risk factors for subsequent UTUC and reduce the effects of selection bias and potential confounders in this observational study, we performed multiple logistic regression analyses. A multivariate analysis was used for the analysis of odds ratios (ORs) and 95% confidence intervals (CIs) for factors with $p < 0.1$ in the univariate analysis. Statistical analyses were performed using the SPSS statistical software (version 26.0; SPSS Japan Inc., Tokyo, Japan). All reported $p$ values were two-sided, and statistical significance was set at $p < 0.05$.

## 3. Results

### 3.1. Patient Characteristics

Of the 3226 patients with NMIBC who were treated with BCG therapy, 2873 were analyzed, and 175 (6.1%) were diagnosed with UTUC during the follow-up period from the initiation of BCG therapy (Table 1). In this patient population, the median period of total observation was 48.0 (7–215) months, and the median period between the first dose of the BCG treatment and diagnosis with UTUC was 27.8 (3–182) months. There were 132 men and 43 women in the UTUC group; the median age was 71.0 (28–86) years when UTUC was confirmed. The location of urothelial carcinoma was the renal pelvis in 33 patients, the ureter in 106 patients, both the pelvis and the ureter in 8 patients, and not described in 27 patients. The pathological findings of the preceding TURBT were Ta or T1 in 124 patients and Tis in 51 patients (Table 1).

**Table 1.** Characteristics of patients with subsequent diagnosis with UTUC after BCG bladder instillation therapy.

| | Subsequent Diagnosis with UTUC (N = 175, 5.4%) | |
|---|---|---|
| Observation duration | Months (range) | 27.8 (0–182) |
| Gender | Male | 132 |
| | Female | 43 |
| Age | Median (range) | 71.0 (28.0–86) |
| Location of UTUC | Renal pelvis | 33 |
| | Ureter | 105 |
| | Renal pelvis + ureter | 8 |
| | Not described | 27 |
| Primary bladder T stage | a or 1 | 124 |
| | CIS | 51 |

UTUC, Urinary tract urothelial carcinoma; BCG, bacillus Calmette-Guerin; T, clinical tumor stage; CIS, carcinoma in situ.

### 3.2. Survival Analyses

Regarding the comparison of survival rates between patients with and without subsequent UTUC, RFS (median 70.5 m vs. 147.3 m, HR 2.552, 95% CI 2.069–3.149, $p < 0.001$), CSS (median 168.3 m vs. 227.4 m, HR 3.434, 95% CI 2.284–5.162, $p < 0.001$), and OS (median 156.0 m vs. 189.6 m, HR 1.485, 95% CI 1.063–2.075, $p = 0.020$) were worse for patients with UTUC (Figure 1).

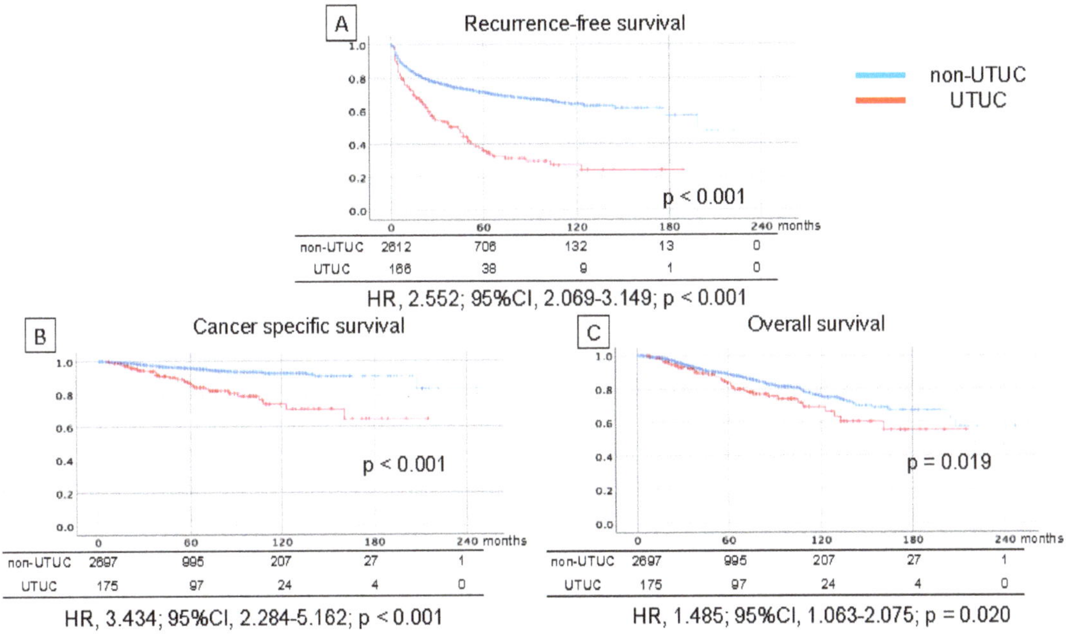


**Figure 1.** Kaplan–Meier curve of recurrence-free survival (**A**), cancer-specific survival (**B**), and overall survival (**C**) in non-muscle-invasive bladder cancer patients with and without subsequent diagnosis of upper tract urothelial carcinoma after intravesical bacillus Calmette-Guerin (BCG) therapy.

### 3.3. Analyses of Predictive Factors for Subsequent UTUC after BCG

We assessed the clinical factors to predict UTUC after BCG therapy and performed a univariate analysis (Table 2). Bladder recurrence after BCG therapy (OR 4.200, 95% CI 3.057–5.770, $p < 0.001$), high-grade carcinoma (OR 2.489, 95% CI 1.211–5.116, $p = 0.009$), multiple bladder tumors (OR 1.980, 95% CI 1.313–2.986, $p = 0.001$), bladder tumor >3 cm (OR 1.815, 95% CI 1.182–2.787, $p = 0.009$), Connaught strain (OR 1.789, 95% CI 1.290–52.183, $p = 0.001$), CIS (OR 1.758, 95% CI 1.252–2.470, $p = 0.002$), female sex (OR 1.577, 95% CI 1.102–2.256, $p = 0.018$), and smoking history (OR 1.504, 95% CI 1.085–2.088, $p = 0.017$) were identified as risk factors for subsequent UTUC (Table 2).

**Table 2.** Clinical predictive factors for subsequent UTUC after BCG bladder instillation therapy in univariate analysis.

| | | | | | 95% CI | | |
|---|---|---|---|---|---|---|---|
| | | **Subsequent UTUC** | **No UTUC** | **OR** | **Lower** | **Upper** | ***p*** |
| Gender | Male:Female | 132:43 | 2236:462 | 1.577 | 1.102 | 2.256 | 0.018 |
| Age | Median (range) | 71.0 (28.0–86.0) | 72.0 (29.0–97.0) | - | - | - | 0.499 |

**Table 2.** *Cont.*

| | | Subsequent UTUC | No UTUC | OR | 95% CI Lower | 95% CI Upper | *p* |
|---|---|---|---|---|---|---|---|
| PS | 0:1 or More | 151:15 | 2317:278 | 0.828 | 0.480 | 1.428 | 0.603 |
| Smoking | Previous and present:No | 77:76 | 1461:959 | 1.504 | 1.085 | 2.088 | 0.017 |
| Grade | High:Low | 159:8 | 2324:291 | 2.489 | 1.211 | 5.116 | 0.009 |
| T | 0 and 1:CIS | 124:51 | 2189:512 | 1.758 | 1.252 | 2.470 | 0.002 |
| Concurrent CIS | Yes:No | 81:37 | 1030:591 | 1.256 | 0.840 | 1.877 | 0.277 |
| Multiplicity | Multiple:Solitary | 130:29 | 1825:806 | 1.980 | 1.313 | 2.986 | 0.001 |
| Maximum diameter (cm) | 3 or more:Less than 3 | 33:68 | 438:1638 | 1.815 | 1.182 | 2.787 | 0.009 |
| Appearance | Papillary:Others | 46:114 | 648:1973 | 1.213 | - | - | 0.259 |
| Variant histology | UC:Others | 5:170 | 63:2619 | 1.206 | - | - | 0.607 |
| Lympho-vascular invasion | Yes:No | 10:164 | 117:2581 | 1.345 | 0.692 | 2.615 | 0.343 |
| BCG strain | Connaught:Tokyo | 59:114 | 603:2086 | 1.789 | 1.290 | 2.183 | 0.001 |
| Incomplete BCG induction therapy | Yes:No | 19:156 | 348:2353 | 0.824 | - | - | 0.485 |
| Maintenance Therapy | Yes:No | 24:151 | 493:2208 | 0.712 | 0.458 | 1.107 | 0.154 |
| Recurrence in bladder | Yes:No | 108:66 | 752:1930 | 4.200 | 3.057 | 5.770 | <0.001 |

UTUC, Upper urinary tract carcinoma; BCG, bacillus Calmette-Guerin; PS, performance status; T, clinical tumor stage; CIS, carcinoma in situ; UC, urothelial carcinoma; OR, odds ratio; CI, confidence interval.

To reduce confounding factors, we applied multiple logistic regression analyses to these risk factors from the univariate analysis. The independent factors for subsequent UTUC after BCG therapy were intravesical recurrence (OR 5.097, 95% CI, 3.225–8.056; $p < 0.001$), Connaught strain (OR 2.211, 95% CI, 1.380–3.543; $p = 0.001$), and multiple tumors at TURBT (OR 1.681, 95% CI, 1.005–2.812; $p = 0.048$) (Table 3).

**Table 3.** Multivariable analysis for elucidating a risk factor of subsequent UTUC by logistic regression analysis.

| Factor | Risk Category | Multivariable OR | 95% CI Lower Limit | 95% CI Upper Limit | *p* |
|---|---|---|---|---|---|
| Gender | female | 1.644 | 0.938 | 2.881 | 0.082 |
| Smoking history | yes | 1.490 | 0.917 | 2.427 | 0.108 |
| Grade | high | 1.405 | 0.591 | 3.342 | 0.442 |
| T | CIS | 1.217 | 0.633 | 2.339 | 0.557 |
| Multiplicity | multiple | 1.681 | 1.005 | 2.812 | 0.048 |
| Tumor diameter | 3 cm≤ | 1.610 | 1.001 | 2.591 | 0.055 |
| Strain | Connaught | 2.211 | 1.380 | 3.543 | 0.001 |
| Intravesical recurrence | yes | 5.097 | 3.225 | 8.056 | <0.001 |

UTUC, Upper urinary tract carcinoma; T, clinical tumor stage; CIS, carcinoma in situ; OR, odds ratio; CI, confidence interval.

## 4. Discussion

In this retrospective, multi-institutional study, subsequent UTUC was detected in 175 (6.1%) patients during the follow-up period after BCG instillation therapy for NMIBC. Patients with UTUC had significantly poorer RFS, CSS, and OS rates than those without UTUC. Intravesical recurrence after the initial BCG therapy, Connaught strain, and tumor multiplicity were associated with a subsequent UTUC diagnosis.

The rate of a subsequent UTUC diagnosis of 6.1% in our retrospective study was equivalent to previous reports about the UTUC rate after total cystectomy [8,9] and after BCG bladder instillation therapy [3,6]. These rates were higher than those of subsequent

UTUC in all UBC populations. To explain this high rate, two possible explanations have been debated, namely, the "field effect" and "tumor seeding" hypotheses, concerning UTUC incidence after UBC [9,15,16]. However, recent next-generation sequencing (NGS) has identified significant clonality between primary UTUC and metachronous UBC in individual patients, strongly suggesting that tumor seeding rather than de novo oncogenesis drives recurrences [17]. This result raises the argument that clinicians might fail to detect UTUC at the time of UBC diagnosis [11]. Further evaluation, such as a paired analysis using NGS between specimens of UBC and subsequent UTUC, might resolve this fundamental question.

Subsequent UTUC had a negative impact on survival rates in our analysis. Although the association between prognosis and UTUC after BCG treatment has been controversial, our study recruited a significant number of patients compared to previous studies [6,11,18–20]. In general, UTUC after BCG therapy might have a clinical risk of a poor survival rate, together with higher tumor stage [20], LVI, and Tis [11] in patients with UBC before BCG treatment. However, our study showed that these risk factors were not associated with a subsequent UTUC diagnosis. Patients with UTUC after cystectomy showed poorer survival rates, which might have directly contributed to poor prognoses in our study.

Connaught strain was identified as a risk factor for subsequent UTUC in our study; however, we do not believe this is a general risk factor worldwide. In terms of bladder recurrence rate, no obvious difference in treatment efficacy among strains has been reported. However, in the patients who were treated with two courses or more [21], general symptoms such as fever were more frequently seen in the patients treated with the Tokyo strain as the first course and the second course compared to patients treated with the Tokyo strain as the first course and the Connaught strain as the second course. This result may help us speculate that the treatment sequence of the same strain affects distant lesions more effectively. The acquired immunology could be boosted by BCG instillation in the Japanese strain because quite a few Japanese were vaccinated with the BCG Tokyo strain during their school days to prevent tuberculosis [22]. This effort might increase treatment efficacy for patients with urothelial carcinoma of the whole urinary tract when they receive BCG bladder instillation therapy.

Our results showed a statistically significant difference in subsequent UTUC in patients with multiple diseases compared to those with a solitary lesion at TURBT before BCG bladder instillation. These multiple diseases might have multifocal abnormalities not only in the bladder but also in the whole urinary tract [7]. Tumor recurrence in the bladder is a major risk factor for tumor progression after BCG [23] and is also reported as a risk factor for subsequent UTUC [6]. There could be some explanations for this relationship. First, early recurrence of UBC after BCG therapy, which was recently defined as "BCG unresponsive" is a high-risk entity in local and systemic disease progression. This risky entity could lead to multifocal disease in the whole urinary tract and could be associated with field effect theory [24]. Second, primary UTUC might already exist during BCG therapy and emit cancer cells into the bladder; this is known as tumor seeding [15]. Schwab et al. [18] reported that 75 patients with positive urine cytology in the absence of visible bladder tumors after a complete response to BCG therapy showed 83% disease recurrence within the urinary tract. Of these recurrences, 20% were detected as UTUCs. Low-grade UTUC might exist more often with negative cytology and be invisible even in recent imaging modality [25,26]. Indeed, patients with UTUC following radical cystectomy have poorer outcomes, even with radical nephroureterectomy.

Although recurrence of UC in the remnant urothelium is a rare event (4%–10% in the upper urinary tract), most patients have an adverse prognosis, despite the absence of distant disease at diagnosis [9]. Therefore, surveillance of the remnant urothelium should be carried out for patients, as it may improve early detection and confer therapeutic benefits. In patients treated with cystectomy, subsequent UC developed mainly in those with risk factors for recurrence [27]. As the number of risk factors has been shown to affect the incidence of subsequent UC, the intensity of surveillance should be based on a risk-adapted strategy [27]. This suggests that patients should be intensively examined for recurrence after BCG therapy, being treated with a non-standard BCG strain, and multiple tumors at TURBT [9]. In patients with risk factors, surveillance, such as urine analysis, urine cytology, and ultrasound sonography, should be conducted at least annually for pan-urothelial disease years after BCG. Diagnostic urethroscopy and cross-sectional imaging of the upper tract should be performed in cases of suspected positivity in screenings [9].

The present study had several limitations. First, since this was a retrospective chart review study, some important information was lacking, such as the reason UTUC was diagnosed, UTUC development after BCG therapy, BCG dose modification details, treatment discontinuation, and concomitant drug therapy. Second, the BCG strain had to be switched from the Connaught strain to the Tokyo strain in the middle of the study because of the supply limitation of the Connaught strain. This change might have affected patient outcomes. Third, the superiority of BCG maintenance therapy since its conception makes it a standard practice today, as it results in less recurrence and improved progression rates in patients with NMIBC with intermediate-to-high risk. However, only 17.9% of the patients in our cohort received it. Finally, there may be a regional bias, and our results may not be generalizable to other populations due to differences in medical practices. Nevertheless, to the best of our knowledge, this is the largest study to report UTUC after BCG instillation therapy in patients with NMIBC.

## 5. Conclusions

Patients with NMIBC with subsequent UTUC showed worse survival outcomes than those without UTUC after BCG bladder instillation therapy. Intravesical recurrence after BCG treatment by Connaught strain and multiple bladder tumors were risk factors for subsequent development of UTUC.

**Supplementary Materials:** The following are available online at https://www.mdpi.com/article/10.3390/cancers15072002/s1, Figure S1: Selection of patients. Out of 3226 patients with non-muscle-invasive bladder cancer treated by bacillus Calmette-Guerin (BCG) bladder instillation therapy, we analyzed 2873 patients who recorded history with or without subsequent diagnosis with upper urinary tract carcinoma.

**Author Contributions:** Conceptualization, K.N., M.M. and M.K.; methodology, K.N., M.M., M.K. and T.H.; software, K.N. and M.K.; formal analysis, K.N. and M.K.; investigation, K.N. and M.K.; resources, K.N.; data curation, M.M., Y.M. (Yumina Muto), Y.S., N.N. (Nobutaka Nishimura), K.I., M.S., S.M., T.Y. (Takahiro Yoneyama), Y.M. (Yoshiaki Matsumura), T.A., T.Y. (Takeshi Yamada), K.M., J.I., N.N. (Naotaka Nishiyama) and R.T.; writing—original draft preparation, K.N. and M.K.; writing—review and editing, K.N. and T.H.; visualization, K.N. and M.K.; supervision, M.M., T.K. (Takashi Kobayashi), T.K. (Takahiro Kojima), H.K., H.N., K.F. and T.H.; Project Administration, K.N. and M.K.; Funding Acquisition, K.N. All authors have read and agreed to the published version of the manuscript.

**Funding:** This research was funded by Grants-in-Aid for Scientific Research, Japan grant numbers 17K11121 and 20K09553.

**Institutional Review Board Statement:** The study was conducted according to the guidelines of the Declaration of Helsinki and approved by the Ethics Committee of Akita University Hospital and that of the other participating institutions (ID 2266). The protocol was approved on 19 July 2019.

**Informed Consent Statement:** Informed consent was obtained from all subjects involved in the study.

**Data Availability Statement:** The data presented in this study are available on request from the corresponding author.

**Acknowledgments:** We are grateful to Yoko Mitobe and Yukiko Sugiyama for their assistance in data collection.

**Conflicts of Interest:** The authors declare no conflict of interest. The funders had no role in the design of the study; in the collection, analyses, or interpretation of data; in the writing of the manuscript; or in the decision to publish the results.

## References


1. Inokuchi, J.; Kuroiwa, K.; Kakehi, Y.; Sugimoto, M.; Tanigawa, T.; Fujimoto, H.; Gotoh, M.; Masumori, N.; Ogawa, O.; Eto, M.; et al. Role of lymph node dissection during radical nephroureterectomy for upper urinary tract urothelial cancer: Multi-institutional large retrospective study JCOG1110A. *World J. Urol.* **2017**, *35*, 1737–1744. [CrossRef] [PubMed]
2. Sfakianos, J.P.; Gul, Z.; Shariat, S.F.; Matin, S.F.; Daneshmand, S.; Plimack, E.; Lerner, S.; Roupret, M.; Pal, S. Genetic Differences between Bladder and Upper Urinary Tract Carcinoma: Implications for Therapy. *Eur. Urol. Oncol.* **2021**, *4*, 170–179. [CrossRef]
3. Roupret, M.; Babjuk, M.; Comperat, E.; Zigeuner, R.; Sylvester, R.J.; Burger, M.; Cowan, N.C.; Gontero, P.; Van Rhijn, B.W.G.; Mostafid, A.H.; et al. European Association of Urology Guidelines on Upper Urinary Tract Urothelial Carcinoma: 2017 Update. *Eur. Urol.* **2018**, *73*, 111–122. [CrossRef]
4. Adibi, M.; Youssef, R.; Shariat, S.F.; Lotan, Y.; Wood, C.G.; Sagalowsky, A.I.; Zigeuner, R.; Montorsi, F.; Bolenz, C.; Margulis, V. Oncological outcomes after radical nephroureterectomy for upper tract urothelial carcinoma: Comparison over the three decades. *Int. J. Urol.* **2012**, *19*, 1060–1066. [CrossRef]
5. Hautmann, R.E.; de Petriconi, R.C.; Pfeiffer, C.; Volkmer, B.G. Radical cystectomy for urothelial carcinoma of the bladder without neoadjuvant or adjuvant therapy: Long-term results in 1100 patients. *Eur. Urol.* **2012**, *61*, 1039–1047. [CrossRef] [PubMed]
6. Nishiyama, N.; Hotta, H.; Takahashi, A.; Yanase, M.; Itoh, N.; Tachiki, H.; Miyao, N.; Matsukawa, M.; Kunishima, Y.; Taguchi, K.; et al. Upper tract urothelial carcinoma following intravesical bacillus Calmette-Guerin therapy for nonmuscle-invasive bladder cancer: Results from a multi-institutional retrospective study. *Urol. Oncol.* **2018**, *36*, 306.e9–306.e15. [CrossRef] [PubMed]
7. Millan-Rodriguez, F.; Chechile-Toniolo, G.; Salvador-Bayarri, J.; Huguet-Perez, J.; Vicente-Rodriguez, J. Upper urinary tract tumors after primary superficial bladder tumors: Prognostic factors and risk groups. *J. Urol.* **2000**, *164*, 1183–1187. [CrossRef]
8. Picozzi, S.; Ricci, C.; Gaeta, M.; Ratti, D.; Macchi, A.; Casellato, S.; Bozzini, G.; Carmignani, L. Upper urinary tract recurrence following radical cystectomy for bladder cancer: A meta-analysis on 13,185 patients. *J. Urol.* **2012**, *188*, 2046–2054. [CrossRef] [PubMed]
9. Gakis, G.; Black, P.C.; Bochner, B.H.; Boorjian, S.A.; Stenzl, A.; Thalmann, G.N.; Kassouf, W. Systematic Review on the Fate of the Remnant Urothelium after Radical Cystectomy. *Eur. Urol.* **2017**, *71*, 545–557. [CrossRef]
10. Seisen, T.; Granger, B.; Colin, P.; Leon, P.; Utard, G.; Renard-Penna, R.; Comperat, E.; Mozer, P.; Cussenot, O.; Shariat, S.F.; et al. A Systematic Review and Meta-analysis of Clinicopathologic Factors Linked to Intravesical Recurrence after Radical Nephroureterectomy to Treat Upper Tract Urothelial Carcinoma. *Eur. Urol.* **2015**, *67*, 1122–1133. [CrossRef]
11. Giannarini, G.; Birkhauser, F.D.; Recker, F.; Thalmann, G.N.; Studer, U.E. Bacillus Calmette-Guerin failure in patients with non-muscle-invasive urothelial carcinoma of the bladder may be due to the urologist's failure to detect urothelial carcinoma of the upper urinary tract and urethra. *Eur. Urol.* **2014**, *65*, 825–831. [CrossRef]
12. Babjuk, M.; Burger, M.; Comperat, E.M.; Gontero, P.; Mostafid, A.H.; Palou, J.; van Rhijn, B.W.G.; Roupret, M.; Shariat, S.F.; Sylvester, R.; et al. European Association of Urology Guidelines on Non-muscle-invasive Bladder Cancer (TaT1 and Carcinoma In Situ)—2019 Update. *Eur. Urol.* **2019**, *76*, 639–657. [CrossRef] [PubMed]
13. Chang, S.S.; Bochner, B.H.; Chou, R.; Dreicer, R.; Kamat, A.M.; Lerner, S.P.; Lotan, Y.; Meeks, J.J.; Michalski, J.M.; Morgan, T.M.; et al. Treatment of Non-Metastatic Muscle-Invasive Bladder Cancer: AUA/ASCO/ASTRO/SUO Guideline. *J. Urol.* **2017**, *198*, 552–559. [CrossRef]
14. Matsumoto, H.; Shiraishi, K.; Azuma, H.; Inoue, K.; Uemura, H.; Eto, M.; Ohyama, C.; Ogawa, O.; Kikuchi, E.; Kitamura, H.; et al. Clinical Practice Guidelines for Bladder Cancer 2019 edition by the Japanese Urological Association: Revision working position paper. *Int. J. Urol.* **2020**, *27*, 362–368. [CrossRef] [PubMed]
15. Habuchi, T.; Takahashi, R.; Yamada, H.; Kakehi, Y.; Sugiyama, T.; Yoshida, O. Metachronous multifocal development of urothelial cancers by intraluminal seeding. *Lancet* **1993**, *342*, 1087–1088. [CrossRef] [PubMed]
16. Jones, T.D.; Wang, M.; Eble, J.N.; MacLennan, G.T.; Lopez-Beltran, A.; Zhang, S.; Cocco, A.; Cheng, L. Molecular evidence supporting field effect in urothelial carcinogenesis. *Clin. Cancer Res.* **2005**, *11*, 6512–6519. [CrossRef] [PubMed]
17. Audenet, F.; Isharwal, S.; Cha, E.K.; Donoghue, M.T.A.; Drill, E.N.; Ostrovnaya, I.; Pietzak, E.J.; Sfakianos, J.P.; Bagrodia, A.; Murugan, P.; et al. Clonal Relatedness and Mutational Differences between Upper Tract and Bladder Urothelial Carcinoma. *Clin. Cancer Res.* **2019**, *25*, 967–976. [CrossRef]
18. Schwalb, M.D.; Herr, H.W.; Sogani, P.C.; Russo, P.; Sheinfeld, J.; Fair, W.R. Positive urinary cytology following a complete response to intravesical bacillus Calmette-Guerin therapy: Pattern of recurrence. *J. Urol.* **1994**, *152*, 382–387. [CrossRef]
19. Merz, V.W.; Marth, D.; Kraft, R.; Ackermann, D.K.; Zingg, E.J.; Studer, U.E. Analysis of early failures after intravesical instillation therapy with bacille Calmette-Guerin for carcinoma in situ of the bladder. *Br. J. Urol.* **1995**, *75*, 180–184. [CrossRef]
20. Herr, H.W. Extravesical tumor relapse in patients with superficial bladder tumors. *J. Clin. Oncol.* **1998**, *16*, 1099–1102. [CrossRef] [PubMed]
21. Niwa, N.; Kikuchi, E.; Matsumoto, K.; Kosaka, T.; Mizuno, R.; Oya, M. Does switching the bacillus Calmette-Guerin strain affect clinical outcome in patients with recurrent non-muscle-invasive bladder cancer after initial bacillus Calmette-Guerin therapy? *Urol. Oncol.* **2018**, *36*, 306.e1–306.e8. [CrossRef]
22. Sakiyama, M.; Kozaki, Y.; Komatsu, T.; Niwa, K.; Suzuki, H.; Ota, M.; Ono, Y.; Miyagawa, Y.; Kiyozumi, T.; Kawana, A. Specificity of tuberculin skin test improved by BCG immunization schedule change in Japan. *J. Infect. Chemother.* **2021**, *27*, 1306–1310. [CrossRef]

23. Kamat, A.M.; Colombel, M.; Sundi, D.; Lamm, D.; Boehle, A.; Brausi, M.; Buckley, R.; Persad, R.; Palou, J.; Soloway, M.; et al. BCG-unresponsive non-muscle-invasive bladder cancer: Recommendations from the IBCG. *Nat. Rev. Urol.* **2017**, *14*, 244–255. [CrossRef] [PubMed]
24. Aragon-Ching, J.B.; Nizam, A.; Henson, D.E. Carcinomas of the Renal Pelvis, Ureters, and Urinary Bladder Share a Carcinogenic Field as Revealed in Epidemiological Analysis of Tumor Registry Data. *Clin. Genitourin. Cancer* **2019**, *17*, 436–442. [CrossRef] [PubMed]
25. Naser-Tavakolian, A.; Ghodoussipour, S.; Djaladat, H. Upper urinary tract recurrence following bladder cancer therapy: A review of surveillance and management. *Curr. Opin. Urol.* **2019**, *29*, 189–197. [CrossRef]
26. Villa, L.; Haddad, M.; Capitanio, U.; Somani, B.K.; Cloutier, J.; Doizi, S.; Salonia, A.; Briganti, A.; Montorsi, F.; Traxer, O. Which Patients with Upper Tract Urothelial Carcinoma Can be Safely Treated with Flexible Ureteroscopy with Holmium:YAG Laser Photoablation? Long-Term Results from a High Volume Institution. *J. Urol.* **2018**, *199*, 66–73. [CrossRef] [PubMed]
27. Volkmer, B.G.; Schnoeller, T.; Kuefer, R.; Gust, K.; Finter, F.; Hautmann, R.E. Upper urinary tract recurrence after radical cystectomy for bladder cancer—Who is at risk? *J. Urol.* **2009**, *182*, 2632–2637. [CrossRef] [PubMed]

*cancers*

MDPI

Article

# Utility of Patient-Reported Symptom and Functional Outcomes to Indicate Recovery after First 90 Days of Radical Cystectomy: A Longitudinal Study

Xin Shelley Wang [1,*], Kelly K. Bree [2], Neema Navai [2], Mona Kamal [1], Shu-En Shen [1], Elizabeth Letona [1], Charles S. Cleeland [1], Qiuling Shi [3] and Vijaya Gottumukkala [4]

1 Department of Symptom Research, The University of Texas MD Anderson Cancer Center, Houston, TX 77030, USA; mkjomaa@mdanderson.org (M.K.); sshen@mdanderson.org (S.-E.S.); egonzalez11@mdanderson.org (E.L.); cscleeland@gmail.com (C.S.C.)
2 Department of Urology, The University of Texas MD Anderson Cancer Center, Houston, TX 77030, USA; kbree@mdanderson.org (K.K.B.); nnavai@mdanderson.org (N.N.)
3 School of Public Health and Management, Chongqing Medical University, Chongqing 400016, China; qshi@cqmu.edu.cn
4 Department of Anesthesiology & Perioperative Medicine, The University of Texas MD Anderson Cancer Center, Houston, TX 77030, USA; vgottumukkala@mdanderson.org
* Correspondence: xswang@mdanderson.org; Tel.: +1-713-745-3504

**Simple Summary:** One barrier to implementing patient-reported outcomes (PROs) during perioperative care for bladder cancer (BLC) patients is the lack of empirical reports of meaningful symptom burden that are associated with postoperative recovery. This study aimed to describe symptom burden and functioning status for 3 months post-radical cystectomy, using a validated disease-specific PRO measure tool, the MD Anderson Symptom Inventory (the MDASI-PeriOp-BLC). We found that the most severe symptom burden at baseline and discharge is associated with poor functional recovery post-radical cystectomy for BLC. These PROs could be used to identify BLC patients at the highest risk for poor functional recovery during the perioperative period. We also found that postoperative functional recovery assessment via PROs is more feasible than an objective performance measure. The completion of MDASI-PeriOp-BLC at preoperative, discharge and end of study was 100%, 79% and 77%, while Timed Up and Go test completion rates were 88%, 54% and 13%, respectively.

**Abstract:** This is a longitudinal prospective study that tracked multiple symptom burden and functioning status for bladder cancer (BLC) patients for 3 months post-radical cystectomy at The University of Texas MD Anderson Cancer Center, using a validated disease-specific patient-reported outcome measure (PROM) tool, the MD Anderson Symptom Inventory (the MDASI-PeriOp-BLC). The feasibility of collecting an objective measure for physical functioning, using "Timed Up & Go test" (TUGT) and PRO scores at baseline, discharge and end of study, was tested. Patients ($n$ = 52) received care under an ERAS pathway. The more severe scores of fatigue, sleep disturbance, distress, drowsiness, frequent urination and urinary urgency at baseline predicted poor functional recovery postoperatively (OR = 1.661, 1.039–2.655, $p$ = 0.034); other more severe symptoms at discharge (pain, fatigue, sleep disturbance, lack of appetite, drowsiness, bloating/abdominal tightness) predicted poor functional recovery (OR = 1.697, 1.114–2.584, $p$ = 0.014) postoperatively. Compliance rates at preoperative, discharge and end of study were 100%, 79% and 77%, while TUGT completion rates were 88%, 54% and 13%, respectively. This prospective study found that more severe symptom burden at baseline and discharge is associated with poor functional recovery post-radical cystectomy for BLC. The collection of PROs is more feasible than using performance measures (TUGT) of function following radical cystectomy.

**Keywords:** postoperative symptoms; functional recovery; patient-reported outcomes (PROs); perioperative care; MD Anderson Symptom Inventory (MDASI); cystectomy

## 1. Introduction

Radical cystectomy, with or without neoadjuvant chemotherapy, is a curative treatment option for recurrent non-muscle invasive or locally advanced stages and a palliative treatment option for the metastatic disease of bladder cancer (BLC) [1–3]. With improved surgical care in patients with cancer, there is increased attention to using patient-reported outcomes (PROs) in perioperative care to improve clinical outcomes [4]. Although the awareness of the role of PROs in many clinical settings is growing, PRO research in BLC postoperative care has not been fully explored.

A systematic review found that only eight BLC randomized clinical trials included PROs between 2014 and 2018, and the quality of reported PROs was found to be inadequate [5]. PROs have been more widely and effectively utilized in other malignancies. PROs have been used as a measure of functional recovery post-thoracic surgery for lung cancer [6] and found to be more feasible to collect compared to objective performance measures in the postoperative setting [7].

While uniform reporting standards for PRO use in BLC are important to fully utilize such data [5], disease/treatment-specific PRO tools are also critical. The MD Anderson Symptom Inventory (MDASI-PeriOpBLC) has been specifically developed and psychometrically validated for assessing symptom burden and functioning in the perioperative period for BLC patients [8], and this was utilized in the current study.

Complications after radical cystectomy occur in 54–80% of patients during the first 90 days after surgery [9–11]. In addition to the typical postoperative symptom burden, such as postoperative pain and fatigue experienced by many surgical patients [12], radical cystectomy represents unique functional recovery challenges [10]. Longitudinally monitoring perioperative symptom burden and functional outcomes might help symptom management in a timely manner, accelerating postoperative recovery, increasing patient satisfaction, and decreasing postoperative complications. To date, there has been relatively little progress in terms of utilization, assessment and understanding of PROs as outcomes in routine postoperative care for BLC [5,13].

To better characterize the role of PROs in the perioperative care of patients with BLC, we conducted a prospective longitudinal study to document patient-reported symptoms and daily functioning before and in the first 90 days after radical cystectomy for BLC. Using a psychometrically validated perioperative version of the MD Anderson Symptom Inventory, MDASI-PeriOp-BLC, we sought to identify patients at the highest risk for poor functional recovery during the perioperative period. We also compared the feasibility of postoperative function recovery assessment via both PROs and an objective performance measure using the "Timed Up & Go test" (TUGT).

## 2. Materials and Methods

### 2.1. Participants

Eligible patients included those at least 18 years old who spoke English, had a diagnosis of BLC and been scheduled for radical cystectomy with curative or palliative intent, had no diagnosis of active psychosis or severe cognitive impairment, understood the study's intent, and were willing to participate. This study was approved by the Institutional Review Board, and written informed consent was obtained from all patients.

### 2.2. Data Collection and PRO Measurement

Eligible BLC patients were enrolled during the preoperative clinic visit at MD Anderson between October 2018 and July 2021. PRO measures were either completed online by the patients or over the phone by a trained study coordinator. In this longitudinal study, we utilized REDCap [14] as the data collection platform to administer the validated MDASI-PeriOp-BLC module (8) and to track electronic PROs (ePROs) perioperatively, daily during hospitalization, on the day of discharge, post-discharge on days 3 and 7, weekly during first 12 weeks after surgery and at the end of the study.

The study coordinator assessed the patients' performance status at the initial assessment using the Eastern Cooperative Oncology Group (ECOG) performance status scale [15], and clinical, demographic and pathological data were collected via chart review. The presence of comorbid conditions according to the Charlson Comorbidity Index [16] was also collected.

The symptom assessment tool MDASI-PeriOp-BLC [8] was developed and psychometrically validated for use in perioperative care for patients with BLC, followed with FDA guidance for development and validation of PRO tools [17]. MDASI-PeriOp-BLC includes 13 common cancer-related symptoms (fatigue, pain, sleep disturbance, poor appetite, distress, drowsiness, dry mouth, shortness of breath, sadness, numbness/tingling, nausea, vomiting, difficulty remembering) and 8 PeriOp-BLC module items (including blood in urine, leaking urine, frequent urination, urinary urgency, constipation, burning with urination, changes in sexual function and stomal problems). Functional status measurement included 6 symptom interferences items (general activity, mood, walk, work, relation with others and enjoyment of life). Patients rated the severity of the symptoms they experienced on a 0–10 numeric rating scale, with 0 meaning no symptom and 10 meaning "as bad as you can imagine". The recall period for measuring symptom severity and functional interferences of the MDASI was the previous 24 h. The cognitive debriefing assessed ease of completion, comprehensibility, acceptability, redundancy, use of the scoring system, item clarification and content domain confirmation of the new instrument.

### 2.3. *Objective Physical Functioning Measure*

The objective-timed performance test (TUGT) [18,19] measures the time it takes a participant to rise from a chair, walk 3 m, turn, walk back and sit down again. TUGT, as a performance outcome measurement tool, was assessed preoperatively, by discharge day, and at the first postoperative outpatient follow-up visit. A final score was recorded as the mean score of two attempts of TUGT at each time point and was categorized into 3 groups (normal ($\leq$10 s), frail (11–20 s) and prolonged (needs further evaluation) (>20 s)), based on the literature [18,19].

### 2.4. *Statistical Analysis*

Descriptive statistics for patients and clinical information were reported as mean, standard deviation, median, minimum and maximum for continuous variables and percent for categorical variables. Loess curves present the most severe symptoms over time during the first 90 days after discharge. We defined the prevalence of moderate symptoms and composite scores of interferences as 5–6 on a 0–10 scale, and severe as $\geq$7 on a 0–10 scale on MDASI-PeriOp-BLC. A composite score selection was based on the 6 items with the highest mean severity score before surgery or at discharge. Composite scores were calculated as the mean of selected items.

An average score of a cluster of the most severe symptoms, both before surgery and at discharge, was calculated. Group-based trajectory modeling (GBTM) [20,21] was used to identify distinct functional recovery trajectories over time with individual PRO items on MDASI Interference [6] and the average score. We generated a two-group model with the goal of simplicity and clinical interpretability that represents two memberships, either high or low symptom burden, over the time period. Individuals with persistently reported high symptom scores are placed in the high-trajectory group, whereas those with persistently low symptom scores are placed in the low-trajectory group. SAS macro PROC TRAJ [22] was used to estimate the trajectory of the MDASI total interference, physical functioning (WAW: work, general activity, walk) and psychological functioning (REM: relation with others, enjoyment of life and mood) scores, using all data collected from discharge to the end of the study.

The strength of the association between the average score of the top severe symptoms before surgery or at discharge and the group of patients with persistent high interference scores (outcome variable) was estimated via multinomial logistic regression models. We used odds ratios to measure the magnitude of the severity of the average score (as a continuous variable) as predictive of poor recovery on patient's daily functioning (composite score of total interferences, physical functioning and psychological functioning) after discharge to 90 days. The covariance in the regression modeling included age (65+ vs. <65 yrs old), CCI (2+ vs. 0–1) and receipt of neoadjuvant chemotherapy (yes vs. no). The association of recovery membership with end of study QoL and end of study health status separately was calculated using paired-sample *t*-tests.

All statistical procedures were performed using SAS statistical software, version 9.4 (SAS Institute, Cary, NC, USA). All $p$ values reported are 2-tailed. Statistical significance was set at $p < 0.05$.

## 3. Results

### 3.1. Participants

A total of 52 patients were enrolled in this longitudinal study and were included in the analysis. Of those patients, two patients withdrew from the study at days 3 and 7, and other patients contributed PROs for the first 90 days (median 89 days, range 32–108 days). Demographic and disease-related characteristics are summarized in Table 1. Enrolled patients were predominantly older adults (mean age 66.5 years old, standard deviation (SD) (9.30)), male (87%) and white (94%). Most of the patients had ECOG performance status of 0–1 (90%), and 27 (52%) had a Charlson Comorbidity Index score of less than or equal to 2. Further, 52% ($n$ = 27) of patients undergoing radical cystectomy had muscle-invasive BLC, while 48% ($n$ = 25) had non-muscle invasive disease. With respect to type of urinary diversion, the majority of patients were treated with an ileal conduit (77%), followed by neobladder (21%) and continent urinary reservoir (2%). The majority (67%) of patients received neoadjuvant chemotherapy, and all received care under an ERAS pathway [23]. The median length of stay was 6 days (IQR: 5–8 days).

**Table 1.** Baseline demographic and clinical characteristics ($n$ = 52).

| Patient Characteristics | Mean (SD) | Median (Range) |
|---|---|---|
| Age, years | 66.52 (9.3) | 66.62 (39.71–86.39) |
| | **Frequency** | **Percent** |
| Age Group | | |
| <65 | 22 | 42.31 |
| ≥65 | 30 | 57.69 |
| Sex | | |
| Men | 45 | 86.54 |
| Women | 7 | 13.46 |
| Race | | |
| White | 49 | 94.23 |
| Others | 3 | 5.77 |
| Ethnicity | | |
| Hispanic or Latino | 2 | 3.85 |
| Not Hispanic | 50 | 96.15 |
| Marital Status | | |
| Married/Partnered | 41 | 78.85 |
| Single/Others | 11 | 21.15 |
| Education | | |
| High School | 11 | 21.15 |
| College | 34 | 65.38 |
| Graduate/Professional training | 7 | 13.46 |

**Table 1.** *Cont.*

| Patient Characteristics | Mean (SD) | Median (Range) |
|---|---|---|
| Job status | | |
| Employed outside the home | 21 | 40.38 |
| Homemaker/Retired | 26 | 50 |
| Medical leave/disable due to illness | 5 | 9.62 |
| ECOG | | |
| Missing | 2 | 3.85 |
| Good (0–1) | 47 | 90.38 |
| Poor (2–3) | 3 | 5.77 |
| Charlson Comorbidity Index | | |
| 0 | 2 | 3.85 |
| 1–2 | 25 | 48.08 |
| 3–4 | 18 | 34.62 |
| 5–7 | 7 | 13.46 |
| Bladder Cancer Stage (Tumor) | | |
| Ta | 2 | 3.85 |
| Tis | 5 | 9.62 |
| T1 | 20 | 38.46 |
| T2 | 21 | 40.38 |
| T3 | 4 | 7.69 |
| Bladder Cancer Stage (Node) | | |
| N0 | 49 | 94.23 |
| N1 | 3 | 5.77 |
| Bladder Cancer Stage (Metastasis) | | |
| M0 | 51 | 98.08 |
| M1 | 1 | 1.92 |
| Under ERAS Pathway | | |
| No | 1 | 1.92 |
| Yes | 51 | 98.08 |
| Intraoperative Complications | | |
| No | 52 | 100 |
| Yes | 0 | 0 |
| Postoperative Complications | | |
| No | 26 | 50 |
| Yes | 26 | 50 |
| Recurrent disease | | |
| No | 42 | 80.77 |
| Yes | 10 | 19.23 |
| Neoadjuvant chemotherapy | | |
| No | 17 | 32.69 |
| Yes | 35 | 67.31 |

### 3.2. *Application of MDASI-PeriOp-BLC*

Patient Compliance

The items with the highest level of missing values post-operatively were frequent urination (9%), urinary urgency (8%), leaking urine (8%), pain or burning with urination (8%) and blood in urine (4%).

Severity and prevalence of symptom burden, health status and quality of life at pre-surgery, discharge and end of study were included.

Figure 1 presents the natural history of perioperative symptom burden in the first 90 days after bladder surgery on MDASI-PeriOp-BLC.

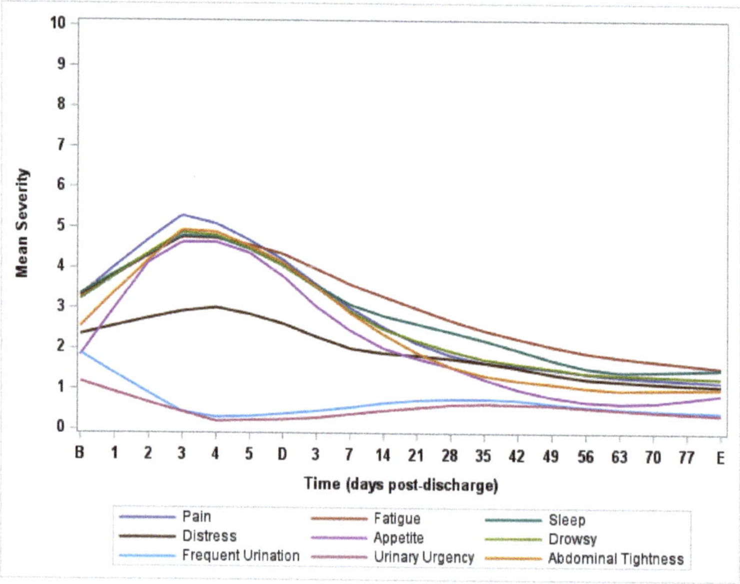


**Figure 1.** Lowess curves of major symptom burden post-radical cystectomy on MDASI-PeriOp-BLC. B = Baseline. D = Discharge. E = End of Study.

Table 2 presents the severity and prevalence of moderate to severe MDASI-PeriOp-BLC symptoms and composite scores of interferences at baseline, discharge and end of study. It also presents the severity of the single-item quality-of-life (SIQOL) question and EQ-5D visual analogue scale (VAS) score to indicate generic health status and quality of life. At baseline, the most severe symptoms were fatigue, disturbed sleep, distress, drowsiness, frequent urination and urinary urgency. At discharge ($n$ = 41), the most severe symptoms were pain, fatigue, disturbed sleep, lack of appetite, drowsiness and bloating/abdominal tightness. At the end of the study ($n$ = 40), an average of 89 days after surgery, few patients reported moderate to severe symptoms. Although we observed significant worsening of multiple general symptom burden items (on MDASI-core symptom items) on discharge day than preoperative (as expected in the immediate postoperative period), there was significant recovery of bladder-cancer-related symptoms on MDASI-PeriOp-BLC module items 90 days post-operatively, with most patients having better MDASI-PeriOp-BLC scores at the end of the study than they did pre-operatively.

**Table 2.** Symptom severity and prevalence of moderate to severe symptoms on MDASI-Periop-BLC, subscales and health status of EQ5D and quality-of-life measure at baseline, discharge and end of study.

| | Baseline | | | | Discharge (6 Days Post-Surgery ± 2 Days) | | | | End of Study (89 Days Post-Surgery ± 2.5 Days) | | | |
|---|---|---|---|---|---|---|---|---|---|---|---|---|
| | *n* | Mean (SD) | % ≥5 | % ≥7 | *n* | Mean (SD) | % ≥5 | % ≥7 | *n* | Mean (SD) | % ≥5 | % ≥7 |
| **Single Item Quality of Life** | 52 | 7.04 (2.82) | 86.54 | 67.31 | 43 | 6.19 (2.39) | 81.40 | 46.51 | 39 | 5.82 (3.01) | 69.23 | 53.85 |
| EQ-5D VAS Score *,† | 52 | 74.33 (22.55) | - | - | 43 | 65.40 (16.19) | - | - | 39 | 78.08 (18.15) | - | - |
| **MDASI-Core Symptom Items** | | | | | | | | | | | | |
| Pain * | 52 | 0.83 (1.45) | 3.85 | 0 | **41** | **3.51 (2.49)** | **29.27** | **17.07** | 40 | 1.45 (2.02) | 10 | 5 |
| Fatigue * | **52** | **1.98 (2.40)** | **15.38** | **5.77** | **41** | **3.54 (2.38)** | **34.15** | **9.76** | 40 | 1.63 (1.85) | 12.5 | 0 |
| Nausea * | 51 | 0.35 (1.25) | 1.92 | 1.92 | 41 | 1.37 (2.22) | 9.76 | 7.32 | 40 | 0.43 (1.43) | 2.5 | 2.5 |
| Disturbed Sleep | **51** | **2.45 (3.01)** | **28.85** | **13.46** | **41** | **3.54 (3.09)** | **31.71** | **21.95** | 40 | 1.78 (2.12) | 15 | 2.5 |
| Distress | **52** | **1.83 (2.17)** | **15.38** | **5.77** | 41 | 2.10 (2.31) | 21.95 | 4.88 | 40 | 1.20 (1.84) | 10 | 2.5 |
| Shortness of Breath | 52 | 0.71 (1.40) | 1.92 | 1.92 | 41 | 1.02 (1.60) | 7.32 | 0 | 40 | 0.75 (1.63) | 5 | 2.5 |
| Difficulty Remembering | 52 | 1.27 (1.73) | 3.85 | 1.92 | 41 | 1.41 (2.40) | 14.63 | 7.32 | 40 | 1.08 (1.59) | 2.5 | 0 |
| Lack of Appetite * | 51 | 0.88 (1.44) | 5.77 | 0 | **41** | **3.66 (2.79)** | **41.46** | **17.07** | 40 | 0.93 (1.61) | 5 | 0 |
| Drowsiness * | **52** | **1.60 (2.36)** | **13.46** | **5.77** | **41** | **3.56 (2.66)** | **34.15** | **9.76** | 40 | 1.18 (1.41) | 2.5 | 0 |
| Dry Mouth * | 52 | 1.13 (2.11) | 11.54 | 3.85 | 41 | 2.80 (2.97( | 24.39 | 14.63 | 40 | 0.73 (1.15) | 2.5 | 0 |
| Sadness | 52 | 1.42 (2.15) | 15.38 | 1.92 | 41 | 1.83 (2.57) | 19.51 | 7.32 | 40 | 1.10 (1.58) | 2.5 | 2.5 |
| Vomiting | 52 | 0.19 (0.99) | 1.92 | 1.92 | 41 | 0.44 (1.38) | 4.88 | 2.44 | 40 | 0.08 (0.27) | 0 | 0 |
| Numbness | 52 | 1.02 (1.84) | 5.77 | 3.85 | 41 | 0.90 (1.61) | 7.32 | 0 | 40 | 1.00 (1.72) | 1 | 0 |
| **MDASI-PeriOpBLC Module Items** | | | | | | | | | | | | |
| Blood in Your Urine † | 52 | 0.96 (2.42) | 7.69 | 5.77 | 37 | 1.57 (2.57) | 17.07 | 7.32 | 39 | 0.08 (0.27) | 0 | 0 |
| Frequent Urination *,† | **52** | **2.83 (2.99)** | **25** | **13.46** | 36 | 0.50 (1.46) | 7.32 | 0 | 39 | 0.51 (1.43) | 5 | 0 |
| Leaking Urine *,† | 52 | 1.44 (2.81) | 13.46 | 11.54 | 36 | 0.47 (1.44) | 4.88 | 0 | 39 | 0.44 (0.94) | 0 | 0 |
| Pain or Burning with Urination *,† | 52 | 0.85 (1.66) | 7.69 | 0 | 36 | 0.03 (0.17) | 0 | 0 | 38 | 0.03 (0.16) | 0 | 0 |
| Urinary Urgency *,† | **52** | **2.42 (2.97)** | **23.08** | **11.54** | 36 | 0.14 (0.54) | 0 | 0 | 39 | 0.49 (1.32) | 5 | 0 |
| Constipation | 52 | 1.04 (1.97) | 7.69 | 1.92 | 41 | 1.90 (2.50) | 19.51 | 9.76 | 40 | 1.15 (1.83) | 12.5 | 0 |
| Diarrhea * | 52 | 0.56 (1.61) | 3.85 | 1.92 | 40 | 1.83 (3.04) | 17.07 | 12.2 | 40 | 0.40 (1.41) | 2.5 | 0 |
| Bloating/Abdominal Tightness *,† | 51 | 0.33 (0.99) | 1.92 | 0 | 41 | **3.10 (2.64)** | **24.39** | **12.2** | **39** | 1.23 (1.72) | 7.5 | 0 |
| Stomal Problems | 51 | 0.25 (0.87) | 1.92 | 0 | 40 | 0.63 (1.21) | 2.44 | 0 | 40 | 0.35 (0.66) | 0 | 0 |
| **MDASI-Interference Items** | | | | | | | | | | | | |
| Walking * | 52 | 1.17 (2.26) | 9.62 | 5.77 | 40 | 3.13 (2.67) | 31.71 | 12.2 | 40 | 1.45 (1.97) | 10 | 2.5 |
| General Activity * | 52 | 2.19 (3.32) | 23.08 | 17.31 | 41 | 5.49 (3.3) | 60.98 | 43.9 | 40 | 2.10 (2.28) | 22.5 | 5 |
| Working * | 52 | 1.58 (2.64) | 19.23 | 5.77 | 40 | 5.00 (4.19) | 51.22 | 48.78 | 40 | 2.20 (2.21) | 15 | 5 |
| Relations with Others * | 52 | 0.88 (2.05) | 9.62 | 1.92 | 41 | 2.02 (2.44) | 19.51 | 7.32 | 40 | 0.93 (1.53) | 7.5 | 0 |
| Enjoyment of Life * | 52 | 1.73 (2.47) | 19.23 | 3.85 | 41 | 3.80 (3.44) | 41.46 | 26.83 | 40 | 1.88 (2.22) | 17.5 | 5 |
| Mood * | 51 | 1.76 (2.45) | 15.38 | 5.77 | 40 | 2.90 (2.82) | 29.27 | 12.2 | 40 | 1.40 (1.75) | 15 | 0 |

**Table 2.** *Cont.*

| | Baseline | | | | Discharge (6 Days Post-Surgery ± 2 Days) | | | | End of Study (89 Days Post-Surgery ± 2.5 Days) | | | |
|---|---|---|---|---|---|---|---|---|---|---|---|---|
| | *n* | Mean (SD) | % ≥5 | % ≥7 | *n* | Mean (SD) | % ≥5 | % ≥7 | *n* | Mean (SD) | % ≥5 | % ≥7 |
| **MDASI Composite Scores** | | | | | | | | | | | | |
| Core * | 52 | 1.20 (1.16) | 0 | 0 | 41 | 2.28 (1.51) | 7.32 | 0 | 40 | 1.02 (1.02) | 0 | 0 |
| Module † | 52 | 1.19 (1.49) | 1.92 | 0 | 41 | 1.30 (1.19) | 2.44 | 0 | 40 | 0.52 (0.59) | 0 | 0 |
| Interference * | 52 | 1.56 (2.23) | 11.54 | 3.85 | 41 | 3.73 (2.47) | 36.59 | 14.63 | 40 | 1.66 (1.64) | 7.5 | 0 |
| WAW * | 52 | 1.65 (2.51) | 15.38 | 5.77 | 41 | 4.59 (2.94) | 53.66 | 24.39 | 40 | 1.92 (1.85) | 5 | 0 |
| REM * | 52 | 1.46 (2.11) | 9.62 | 1.92 | 41 | 2.89 (2.45) | 21.95 | 9.76 | 40 | 1.40 (1.67) | 10 | 0 |
| Most severe 6 Symptoms at Baseline † | 52 | 2.22 (2.14) | 13.46 | 5.77 | 41 | 2.17 (1.61) | 7.32 | 0 | 40 | 1.07 (1.26) | 0 | 0 |
| Most severe 6 Symptoms at Discharge * | 52 | 1.34 (1.41) | 1.92 | 0 | 41 | 3.48 (2.11) | 24.39 | 4.88 | 40 | 1.36 (1.43) | 2.50 | 0 |

The most severe 6 symptoms at baseline composite score includes fatigue, disturbed sleep, distress, drowsiness, frequent urination, and urinary urgency. The most severe 6 symptoms at discharge composite score includes pain, fatigue, disturbed sleep, lack of appetite, drowsiness, and bloating/abdominal tightness. Bolded rows indicate top symptoms at given timepoint. * Significant difference in mean between baseline and discharge. † Significant difference in mean between end of study and baseline.

Table 2 also presents that 37% of patients reported moderate to severe levels of functional interferences at 5 or greater, or 7 or greater on the total scores of MDASI Interferences (general activity, mood, work, walking, relations with others and enjoyment of life) at discharge, which was much higher than pre-surgery (11.5%) and 90-days post-surgery (5.5%).

### 3.3. Trajectory Membership of Postoperative Symptom Functioning Recovery

Figure 2 presents the results from trajectory analysis; a group of 66.7% of patients reported persistently higher symptom interferences over time post-surgery on the MDASI total interference composite score.

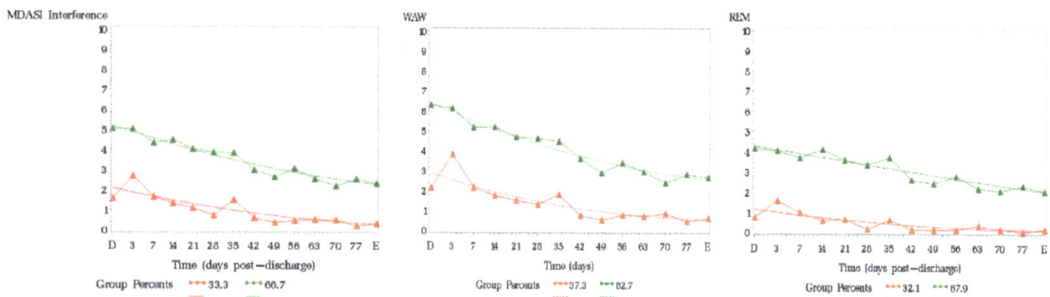


**Figure 2.** Functioning recovery trend by trajectory of membership of MDASI-Interferences. D = Discharge. E = End of Study.

The group with high interference scores had a significant association with patient-reported health status on EQ5D5L (but not with SIQOL) at the first postoperative follow-up ($n$ = 36, mean 41.83 days post-surgery) in the domains of usual activities ($p$ < 0.0001), self-care ($p$ = 0.0015), pain/discomfort ($p$ = 0.0337) and VAS of health status ($p$ = 0.001).

### 3.4. Impact of Symptom Severity on Postoperative Function Recovery

Regression analysis in Table 3 shows that the composite score of the six most severe symptoms at baseline (disturbed sleep, fatigue, distress, drowsiness, frequent urination, urinary urgency) were significant predictors of the interference symptom trajectory group in the univariate model (OR 1.661, 95% CI 1.039–2.655, $p = 0.0339$). Table 3 also presents the composite scores of the six most severe symptoms at discharge (fatigue, pain, disturbed sleep, lack of appetite, drowsiness and bloating/abdominal tightness) that were significantly associated with high symptom trajectory group membership in the multivariate model (OR 1.697, 95% CI 1.114–2.584, $p = 0.014$). Both models were controlled for age, CCI and neoadjuvant chemotherapy status.

**Table 3.** Multivariate analysis of the composite score of most severe symptoms at preoperation and by discharge predicting slow post-operative functioning recovery on MDASI Interference trajectory group membership.

| Outcomes on MDASI | Effect | Odds Ratio | 95% CI | *p*-Value |
|---|---|---|---|---|
| High vs. Low membership of total Interference Composite | Severity of Top 6 Baseline Composite [a] (0–10 scale) | **1.661** | **1.039–2.655** | **0.0339** |
| | CCI (>2 vs. ≤2) | 0.353 | 0.056–2.246 | 0.2701 |
| | Age (≥65 vs. <65) | 1.504 | 0.244–9.255 | 0.6600 |
| | Received Neoadjuvant Chemotherapy (Yes vs. No) | 1.598 | 0.410–6.235 | 0.4998 |
| High vs. Low membership of total Interference Composite | Severity of Top 6 Discharge Composite [b] (0–10 scale) | **1.697** | **1.114–2.584** | **0.0137** |
| | CCI (>2 vs. ≤2) | 1.369 | 0.157–11.960 | 0.7766 |
| | Age (≥65 vs. <65) | 0.564 | 0.064–4.939 | 0.6052 |
| | Received Neoadjuvant Chemotherapy (Yes vs. No) | 1.233 | 0.272–5.596 | 0.7863 |
| High vs. Low membership of WAW [c] Composite | Top 6 Baseline Composite [a] | **1.494** | **1.001–2.230** | **0.0494** |
| | CCI (>2 vs. ≤2) | **0.148** | **0.023–0.940** | **0.0428** |
| | Age (≥65 vs. <65) | 1.858 | 0.316–10.909 | 0.4928 |
| | Received Neoadjuvant Chemotherapy (Yes vs. No) | 1.102 | 0.282–4.302 | 0.8891 |
| High vs. Low membership of WAW [c] Composite | Top 6 Discharge Composite [b] | **1.654** | **1.093–2.503** | **0.0172** |
| | CCI (>2 vs. ≤2) | 0.316 | 0.043–2.336 | 0.2592 |
| | Age (≥65 vs. <65) | 0.839 | 0.117–6.032 | 0.8611 |
| | Received Neoadjuvant Chemotherapy (Yes vs. No) | 0.628 | 0.130–3.036 | 0.5631 |
| High vs. Low membership of REM [d] Composite | Severity of Top 6 Baseline Composite [a] (0–10 scale) | **1.617** | **1.025–2.552** | **0.0388** |
| | CCI (>2 vs. ≤2) | 0.451 | 0.076–2.691 | 0.3822 |
| | Age (≥65 vs. <65) | 1.864 | 0.323–10.744 | 0.4861 |
| | Received Neoadjuvant Chemotherapy (Yes vs. No) | 1.044 | 0.266–4.092 | 0.9511 |
| High vs. Low membership of REM [d] Composite | Severity of Top 6 Discharge Composite [b] (0–10 scale) | **1.573** | **1.061–2.331** | **0.0242** |
| | CCI (>2 vs. ≤2) | 1.681 | 0.219–12.883 | 0.6171 |
| | Age (≥65 vs. <65) | 0.726 | 0.095–5.553 | 0.7577 |
| | Received Neoadjuvant Chemotherapy (Yes vs. No) | 0.799 | 0.181–3.534 | 0.7676 |

[a] Top 6 symptoms at baseline include disturbed sleep, fatigue, distress, drowsiness, frequent urination, and urinary urgency. [b] Top 6 symptoms at discharge include lack of appetite, drowsiness, fatigue, disturbed sleep, pain, and bloated/abdominal tightness. [c] WAW includes walking, general activity, and working. [d] REM includes relations with others, enjoyment of life, and mood.

### 3.5. Compliance with Objective and Subjective Measures of Physical Functioning

For the functional assessment, the data available for both PROs on MDASI Interferences and TUGT were recorded. Compliance rates at preoperative, discharge and end of study of MDASI Interferences were 100%, 79% and 77%, while TUGT completion rates were 88%, 54% and 13%, respectively.

Most patients at the baseline and end of study time points did not complete the TUGT test because they completed PRO assessments remotely but were not able to perform TUGT. At the discharge time point, most of the patients who did not complete the TUGT test declined to complete the test.

Spearman correlation showed that the severity of the patient-reported "Walking" interference item on MDASI-I was significantly associated with prolonged TUGT score at baseline (r = 0.3256, $p$ = 0.0273).

## 4. Discussion

This longitudinal study demonstrated the feasibility of utilizing PROs for measuring both symptom and functional outcomes via the MDASI-PeriOp-BLC module, before and during the first 90 days after radical cystectomy. The current study demonstrated that more severe symptom burden, both at baseline and discharge, is predictive of poor functional recovery after radical cystectomy for BLC. However, our study also highlights that despite severe symptom scores in the pre- and immediate postoperative setting, by 3 months post-operatively, many patients had improved QOL and functional status compared to their baseline. Compared to the objective-timed performance test on physical functioning measure (TUGT), completion of PROs (MDASI Interference) was more feasible for monitoring functional status in the BLC postoperative setting.

PRO data are essential to implement personalized treatment plans after cystectomy and can be helpful in predicting outcomes and treatment side effects. Somani et al. [24] previously showed no difference in QoL of patients pre- and post-cystectomy with neobladders, using European Organization for Research and Treatment of Cancer QLQ-C30 and Satisfaction With Life Scale. In a study assessing quality of life (QOL) in BLC patients, two-thirds of the respondents reported at least one HRQOL problem, with mobility issues being most commonly reported, followed by pain and discomfort [25]. Further, in a study among patients undergoing GYN surgery, we reported critical disease-/treatment-specific symptoms (abdominal bloating and cramping) at discharge that were found to be significantly relevant to assessing the risk of grade 2 or higher complications 30 days post-laparotomy [26]. In the current study, we elucidated the recovery characteristics by using a validated PROM tool, MDASI-Periop-BLC, defining the symptom cluster at critical timepoints that are clinically meaningful during the postoperative period after cystectomy. Nevertheless, as shown in Tables 2 and 3, frequent urination and urinary urgency before cystectomy, and the post-cystectomy symptoms of bloating/abdominal tightness at discharge, are time-sensitive and clinically meaningful to determine the surgical recovery, while other major general symptom burden on MDASI core items (pain, fatigue, sleep disturbance, drowsiness) is also critical to include for monitoring and identifying individuals who might experience a less-ideal functioning recovery journey 3 months after surgery.

Measuring functional recovery is an important part of postoperative care. Our study indicated that 37% of patients reported moderate to severe levels of symptom interference at discharge (based upon 5 or greater on MDASI Interferences), although with a trend of improvement over time. As a marker of impaired recovery, we found that MDADI Interference scores were significantly related to multiple domains of health status on EQ5D5L at the first postoperative follow-up clinic visit. As with our earlier reports in patients undergoing thoracic and abdominal procedures [6,7,26], this study adds validity to the use of the interference item subscale of MDASI (MDASI-I) as a PRO functional measure in postoperative care of patients' post-radical cystectomy.

In a previous study among patients undergoing laparotomy for gynecologic tumors [7], we reported that MDASI-I could be used as a surrogate or potential substitution for TUGT to predict patient's physical functioning, both pre- and postoperatively. Consistent with our previous impressions from the gynecologic surgery cohort, this study again confirms that the subjective functioning measure via PROs is a valid and much easier measure of functional status than an in-person objective measure with TUGT. Similar to other studies on TUGT, in which prolonged time was associated with the need for physical assistance [27–29], in this study, we found that the severity of the patient-reported "Walking" interference item on MDASI-I was significantly associated with a prolonged TUGT score at baseline (r = 0.3256, $p$ = 0.0273).

Currently, increased research supports patient empowerment through the integration of real-time PRO monitoring in postoperative care. For high-risk individuals, PRO assessment coupled with triage responsive interventions have the potential to improve the quality of perioperative symptom management in cancer patients [30,31] and enhance functional recovery after surgery [6,32–36]. Our study provides evidence to support future studies on the effectiveness of PROs after radical cystectomy to improve functioning recovery in this cohort of patients.

This study has several limitations. This real-world study is a single-institution study performed at a large tertiary care cancer center, where the majority of the patients are non-Hispanic White males. A further study should be conducted with a more diverse patient population to confirm our findings. Additionally, nearly half of the patients undergoing radical cystectomy in this cohort underwent surgery for non-muscle invasive disease. These patients had very high-risk non-muscle invasive disease, often associated with lymphovascular invasion, variant histology, etc., and many received neoadjuvant chemotherapy. We expect that the use of PROs in the assessment of functional recovery after extirpative surgery is equally as effective in these patients as their muscle-invasive counterparts. Finally, the pre-selected cutoff points for the presented moderate to severe symptoms at 5+ and 7+ on a 0–10 scale of severity were not determined individually, which should be confirmed in a future study for their clinical meaningfulness in BLC postoperative care.

## 5. Conclusions

In conclusion, the monitoring of perioperative symptoms and function using PROs is feasible during the perioperative period after radical cystectomy. Functional status assessment using PROs correlated with functional performance, as measured by more time- and resource-intensive objective measurements such as the TUGT. The impact of routine PRO monitoring on improving postoperative recovery warrants further investigation.

**Author Contributions:** Conceptualization, X.S.W., K.K.B., N.N., C.S.C., Q.S. and V.G.; methodology, X.S.W., M.K., S.-E.S., C.S.C., Q.S. and V.G.; data analysis, S.-E.S. and Q.S.; investigation, K.K.B., N.N. and X.S.W.; resources, X.S.W. and N.N.; data curation, E.L.; writing—original draft preparation, X.S.W., K.K.B., N.N., M.K. and S.-E.S.; writing—review and editing, all; funding acquisition, X.S.W. All authors have read and agreed to the published version of the manuscript.

**Funding:** This study was supported by a grant from the National Institutes of Health/National Cancer Institute, R01CA205146, to XS Wang.

**Institutional Review Board Statement:** The study was conducted in accordance with the Declaration of Helsinki and approved by the Institutional Review Board (or Ethics Committee) of The University of Texas-MD Anderson Cancer Center (protocol code BS99-094 IRB, approved on 5/8/2019).

**Informed Consent Statement:** Informed consent was obtained from all subjects involved in the study.

**Data Availability Statement:** Any data needed to evaluate the conclusions in the paper are present within the paper. Please contact with corresponding author if a further question for research data, which will be considered under institutional ethics policy.

**Acknowledgments:** The authors acknowledge Araceli Garcia-Gonzalez, for her administrative support to the project.

**Conflicts of Interest:** The authors declare no conflict of interest.

## References


1. Grossman, H.B.; Natale, R.B.; Tangen, C.M.; Speights, V.O.; Vogelzang, N.J.; Trump, D.L.; de Vere White, R.W.; Sarosdy, M.F.; Wood, D.P., Jr.; Raghavan, D.; et al. Neoadjuvant Chemotherapy plus Cystectomy Compared with Cystectomy Alone for Locally Advanced Bladder Cancer. *N. Engl. J. Med.* **2003**, *349*, 859–866. [CrossRef]
2. NCCN. NCCN Clinical Practice Guidelines in Oncology-Bladder Cancer, Version 3.2019. 2019. Available online: https://www.nccn.org/professionals/physician_gls/pdf/bladder.pdf (accessed on 1 January 2019).
3. Vashistha, V.; Quinn, D.I.; Dorff, T.B.; Daneshmand, S. Current and recent clinical trials for perioperative systemic therapy for muscle invasive bladder cancer: A systematic review. *BMC Cancer* **2014**, *14*, 966. [CrossRef] [PubMed]
4. Wang, X.S.; Shi, Q.; Williams, L.A.; Cleeland, C.S.; Garcia-Gonzalez, A.; Chen, T.-Y.; Shahid, D.R.; Ramirez, P.T.; Iniesta, M.D.; Siverand, A.M.; et al. Validation and application of a module of the MD Anderson Symptom Inventory for measuring perioperative symptom burden in patients with gynecologic cancer (the MDASI-PeriOp-GYN). *Gynecol. Oncol.* **2019**, *152*, 492–500. [CrossRef] [PubMed]
5. Van Hemelrijck, M.; Sparano, F.; Josephs, D.; Sprangers, M.; Cottone, F.; Efficace, F. Patient-reported outcomes in randomised clinical trials of bladder cancer: An updated systematic review. *BMC Urol.* **2019**, *19*, 86. [CrossRef] [PubMed]
6. Shi, Q.; Wang, X.S.; Vaporciyan, A.A.; Rice, D.C.; Popat, K.U.; Cleeland, C.S. Patient-Reported Symptom Interference as a Measure of Postsurgery Functional Recovery in Lung Cancer. *J. Pain Symptom Manag.* **2016**, *52*, 822–831. [CrossRef]
7. Wang, X.S.; Kamal, M.; Chen, T.H.; Shi, Q.; Garcia-Gonzalez, A.; Iniesta, M.D.; Cleeland, C.S.; Gottumukkala, V.; Meyer, L.A. Assessment of physical function by subjective and objective methods in patients undergoing open gynecologic surgery. *Gynecol. Oncol.* **2021**, *161*, 83–88. [CrossRef] [PubMed]
8. Kamal, M.; Navai, N.; Bree, K.K.; Williams, L.A.; Cleeland, C.S.; Shen, S.-E.; Wang, X.S. Validation and Application of MD Anderson Symptom Inventory Module for Patients with Bladder Cancer in the Perioperative Setting. *Cancers* **2022**, *14*, 3896. [CrossRef]
9. Djaladat, H.; Katebian, B.; Bazargani, S.T.; Miranda, G.; Cai, J.; Schuckman, A.K.; Daneshmand, S. 90-Day complication rate in patients undergoing radical cystectomy with enhanced recovery protocol: A prospective cohort study. *World J. Urol.* **2017**, *35*, 907–911. [CrossRef]
10. Shabsigh, A.; Korets, R.; Vora, K.C.; Brooks, C.M.; Cronin, A.M.; Savage, C.; Raj, G.; Bochner, B.H.; Dalbagni, G.; Herr, H.W.; et al. Defining Early Morbidity of Radical Cystectomy for Patients with Bladder Cancer Using a Standardized Reporting Methodology. *Eur. Urol.* **2009**, *55*, 164–176. [CrossRef]
11. Yuh, B.E.; Nazmy, M.; Ruel, N.H.; Jankowski, J.T.; Menchaca, A.R.; Torrey, R.R.; Linehan, J.A.; Lau, C.S.; Chan, K.G.; Wilson, T.G. Standardized Analysis of Frequency and Severity of Complications After Robot-assisted Radical Cystectomy. *Eur. Urol.* **2012**, *62*, 806–813. [CrossRef]
12. Westerman, M.E.; Kokorovic, A.; Wang, X.S.; Lim, A.; Garcia-Gonzalez, A.; Seif, M.; Wang, R.; Kamat, A.M.; Dinney, C.P.; Navai, N. Radical Cystectomy and Perioperative Sexual Function: A Cross-Sectional Analysis. *J. Sex. Med.* **2020**, *17*, 1995–2004. [CrossRef] [PubMed]
13. Wong, R.L.; Morgans, A.K. Integration of Patient Reported Outcomes in Drug Development in Genitourinary Cancers. *Curr. Oncol. Rep.* **2020**, *22*, 21. [CrossRef] [PubMed]
14. Harris, P.A.; Taylor, R.; Thielke, R.; Payne, J.; Gonzalez, N.; Conde, J.G. Research electronic data capture (REDCap)—A metadata-driven methodology and workflow process for providing translational research informatics support. *J. Biomed. Inform.* **2009**, *42*, 377–381. [CrossRef] [PubMed]
15. Oken, M.M.; Creech, R.H.; Tormey, D.C.; Horton, J.; Davis, T.E.; McFadden, E.T.; Carbone, P.P. Toxicity and response criteria of the Eastern Cooperative Oncology Group. *Am. J. Clin. Oncol.* **1982**, *5*, 649–655. [CrossRef]
16. Charlson, M.E.; Pompei, P.; Ales, K.L.; MacKenzie, C.R. A new method of classifying prognostic comorbidity in longitudinal studies: Development and validation. *J. Chronic Dis.* **1987**, *40*, 373–383. [CrossRef]
17. U.S. Food and Drug Administration. Guidance for Industry-Patient-Reported Outcome Measures: Use in Medical Product Development to Support Labeling Claims 2009. Available online: https://www.fda.gov/media/77832/download (accessed on 1 October 2009).
18. Mathias, S.; Nayak, U.S.; Isaacs, B. Balance in elderly patients: The "get-up and go" test. *Arch. Phys. Med. Rehabil.* **1986**, *67*, 387–389.
19. Thompson, M.; Medley, A. Performance of Community Dwelling Elderly on the Timed Up and Go Test. *Phys. Occup. Ther. Geriatr.* **1995**, *13*, 17–30. [CrossRef]
20. Shi, Q.; Mendoza, T.R.; Gunn, G.B.; Wang, X.S.; Rosenthal, D.I.; Cleeland, C.S. Using group-based trajectory modeling to examine heterogeneity of symptom burden in patients with head and neck cancer undergoing aggressive non-surgical therapy. *Qual. Life Res.* **2013**, *22*, 2331–2339. [CrossRef]
21. Nagin, D.S.; Tremblay, R.E. Analyzing developmental trajectories of distinct but related behaviors: A group-based method. *Psychol. Methods* **2001**, *6*, 18–34. [CrossRef]
22. Jones, B.L. TRAJ. Available online: https://www.andrew.cmu.edu/user/bjones/ (accessed on 21 May 2023).
23. Cerantola, Y.; Valerio, M.; Persson, B.; Jichlinski, P.; Ljungqvist, O.; Hubner, M.; Kassouf, W.; Muller, S.; Baldini, G.; Carli, F.; et al. Guidelines for perioperative care after radical cystectomy for bladder cancer: Enhanced Recovery After Surgery (ERAS®) society recommendations. *Clin. Nutr.* **2013**, *32*, 879–887. [CrossRef]

24. Somani, B.K.; Gimlin, D.; Fayers, P.; N'Dow, J. Quality of Life and Body Image for Bladder Cancer Patients Undergoing Radical Cystectomy and Urinary Diversion—A Prospective Cohort Study with a Systematic Review of Literature. *Urology* **2009**, *74*, 1138–1143. [CrossRef]
25. Catto, J.W.; Downing, A.; Mason, S.; Wright, P.; Absolom, K.; Bottomley, S.; Hounsome, L.; Hussain, S.; Varughese, M.; Raw, C.; et al. Quality of Life After Bladder Cancer: A Cross-sectional Survey of Patient-reported Outcomes. *Eur. Urol.* **2021**, *79*, 621–632. [CrossRef]
26. Wang, X.S.; Ramirez, P.T.; Shi, Q.; Kamal, M.; Garcia-Gonzalez, A.; Iniesta, M.D.; Cleeland, C.S.; Meyer, L.A. Patient-reported symptoms at discharge and risk of complications after gynecologic surgery. *Int. J. Gynecol. Cancer* **2022**, *33*, 271–277. [CrossRef]
27. Bischoff, H.A.; Stähelin, H.B.; Monsch, A.U.; Iversen, M.D.; Weyh, A.; von Dechend, M.; Akos, R.; Conzelmann, M.; Dick, W.; Theiler, R. Identifying a cut-off point for normal mobility: A comparison of the timed 'up and go' test in community-dwelling and institutionalised elderly women. *Age Ageing* **2003**, *32*, 315–320. [CrossRef]
28. Mousa, S.M.; Rasheedy, D.; El-Sorady, K.E.; Mortagy, A.K. Beyond mobility assessment: Timed up and go test and its relationship to osteoporosis and fracture risk. *J. Clin. Gerontol. Geriatr.* **2016**, *7*, 48–52. [CrossRef]
29. Podsiadlo, D.; Richardson, S. The Timed "Up & Go": A Test of Basic Functional Mobility for Frail Elderly Persons. *J. Am. Geriatr. Soc.* **1991**, *39*, 142–148. [CrossRef] [PubMed]
30. Schmidt, M.; Eckardt, R.; Scholtz, K.; Neuner, B.; von Dossow-Hanfstingl, V.; Sehouli, J.; Stief, C.G.; Wernecke, K.-D.; Spies, C.D. PERATECS Group Patient Empowerment Improved Perioperative Quality of Care in Cancer Patients Aged ≥ 65 Years—A Randomized Controlled Trial. *PLoS ONE* **2015**, *10*, e0137824. [CrossRef] [PubMed]
31. Dai, W.; Feng, W.; Zhang, Y.; Wang, X.S.; Liu, Y.; Pompili, C.; Xu, W.; Xie, S.; Wang, Y.; Liao, J.; et al. Patient-Reported Outcome-Based Symptom Management Versus Usual Care After Lung Cancer Surgery: A Multicenter Randomized Controlled Trial. *J. Clin. Oncol.* **2022**, *40*, 988–996. [CrossRef] [PubMed]
32. Cleeland, C.S.; Wang, X.S.; Shi, Q.; Mendoza, T.R.; Wright, S.L.; Berry, M.D.; Malveaux, D.; Shah, P.K.; Gning, I.; Hofstetter, W.L.; et al. Automated Symptom Alerts Reduce Postoperative Symptom Severity After Cancer Surgery: A Randomized Controlled Clinical Trial. *J. Clin. Oncol.* **2011**, *29*, 994–1000. [CrossRef]
33. Fagundes, C.P.; Shi, Q.; Vaporciyan, A.A.; Rice, D.C.; Popat, K.U.; Cleeland, C.S.; Wang, X.S. Symptom recovery after thoracic surgery: Measuring patient-reported outcomes with the MD Anderson Symptom Inventory. *J. Thorac. Cardiovasc. Surg.* **2015**, *150*, 613–619.e2. [CrossRef] [PubMed]
34. Adamina, M.; Gié, O.; Demartines, N.; Ris, F. Contemporary perioperative care strategies. *Br. J. Surg.* **2013**, *100*, 38–54. [CrossRef] [PubMed]
35. Sibbern, T.; Sellevold, V.B.; Steindal, S.A.; Dale, C.; Watt-Watson, J.; Dihle, A. Patients' experiences of enhanced recovery after surgery: A systematic review of qualitative studies. *J. Clin. Nurs.* **2017**, *26*, 1172–1188. [CrossRef] [PubMed]
36. Meyer, L.A.; Nick, A.M.; Shi, Q.; Wang, X.S.; Williams, L.; Brock, T.; Iniesta, M.D.; Rangel, K.; Lu, K.H.; Ramirez, P.T. Perioperative trajectory of patient reported symptoms: A pilot study in gynecologic oncology patients. *Gynecol. Oncol.* **2015**, *136*, 440–445. [CrossRef] [PubMed]

*cancers*

MDPI

*Article*

# Increasing Patient Safety and Treatment Quality by Using Intraoperative MRI for Organ-Preserving Tumor Resection and High-Dose Rate Brachytherapy in Children with Bladder/Prostate and Perianal Rhabdomyosarcoma

**Andreas Schmidt [1,*,†], Constantin Roder [2,3,†], Franziska Eckert [4,5], David Baumann [4], Maximilian Niyazi [4], Frank Fideler [6], Ulrike Ernemann [3,7], Marcos Tatagiba [2,3], Jürgen Schäfer [8], Cristian Urla [1], Simon Scherer [1], Jörg Fuchs [1,9], Frank Paulsen [4,‡] and Benjamin Bender [3,7,‡]**

1 Department of Pediatric Surgery and Pediatric Urology, University Children's Hospital, Eberhard Karls University Tuebingen, 72076 Tuebingen, Germany
2 Department of Neurosurgery, University Hospital, Eberhard Karls University Tuebingen, 72076 Tuebingen, Germany; constantin.roder@med.uni-tuebingen.de (C.R.)
3 Center for Neuro-Oncology, Comprehensive Cancer Center Tuebingen-Stuttgart, University Hospital, Eberhard Karls University Tuebingen, 72070 Tuebingen, Germany; benjamin.bender@med.uni-tuebingen.de (B.B.)
4 Department of Radiation Oncology, University Hospital, Eberhard Karls University Tuebingen, 72076 Tuebingen, Germany; frank.paulsen@med.uni-tuebingen.de (F.P.)
5 Department of Radiation Oncology, AKH, Comprehensive Cancer Center Vienna, Medical University Vienna, 1090 Vienna, Austria
6 Department of Anesthesiology and Intensive Care Medicine, University Hospital, Eberhard Karls University Tuebingen, 72076 Tuebingen, Germany
7 Department of Diagnostic and Interventional Neuroradiology, University Hospital, Eberhard Karls University Tuebingen, 72076 Tuebingen, Germany
8 Department of Pediatric Radiology, University Hospital, Eberhard Karls University Tuebingen, 72076 Tuebingen, Germany
9 Center for Pediatric Oncology, Comprehensive Cancer Center Tuebingen-Stuttgart, University Hospital, Eberhard Karls University Tuebingen, 72070 Tuebingen, Germany
* Correspondence: andreas.schmidt@med.uni-tuebingen.de
† These authors contributed equally to this work.
‡ These authors contributed equally to this work.

**Simple Summary:** The combination treatment of organ-preserving tumor resection and brachytherapy in children with bladder/prostate and perianal rhabdomyosarcoma can reduce therapy-associated side effects while maintaining excellent oncological outcome. This highly individualized hybrid treatment concept poses specific challenges for all clinicians involved in the local treatment. The aim of this study was to determine whether the use of an intraoperative MRI can improve the clinical workflow. These findings may have a positive impact on the treatment quality and patient safety of children with bladder/prostate and perianal RMS.

**Abstract:** In children with bladder/prostate (BP) and perianal rhabdomyosarcoma (RMS), we use a hybrid treatment concept for those suitable, combining organ-preserving tumor resection and high-dose rate brachytherapy (HDR-BT). This treatment concept has been shown to improve outcomes. However, it is associated with specific challenges for the clinicians. The exact position of the tubes for BT is a prerequisite for precise radiotherapy. It can finally be determined only with an MRI or CT scan. We evaluated the use of an intraoperative MRI (iMRI) to control the position of the BT tubes and for radiotherapy planning in all patients with BP and perianal RMS who received the above-mentioned combination therapy in our department since January 2021. iMRI was used in 12 children. All tubes were clearly localized. No adverse events occurred. In all 12 children, radiotherapy could be started on time. In a historical cohort without iMRI, this was not possible in 3 out of 20 children. The use of iMRI in children with BP and perianal RMS improved patient safety and treatment quality. This technology has proven to be successful for the patient population we have defined and has become a standard procedure in our institution.

**Keywords:** rhabdomyosarcoma; bladder/prostate; perianal; brachytherapy; intraoperative MRI

## 1. Introduction

Rhabdomyosarcoma (RMS) is the most common soft tissue sarcoma in children [1]. Multimodal treatment concepts consist of chemotherapy followed by local therapy comprising surgery and/or radiotherapy [2,3]. International and interdisciplinary efforts have resulted in an overall 5-year survival rate higher than 70% [1]. Among the disciplines involved in the treatment, a focus has been put on the prevention of therapy-associated complications [4–6], including organ preservation and the reduction of radiotherapy-associated sequelae.

The combination treatment of organ-preserving tumor resection and brachytherapy (BT) has been established for selected tumor sites when certain criteria are met [7–9]. For RMS in the area of the urinary bladder/prostate (BP), this procedure can be considered a standard procedure nowadays [10,11]. In perianal RMS, which is less frequent but has a considerably worse prognosis, this kind of treatment is also used in individual cases [8,9,12,13]. BT offers advantages compared to conventional radiation modalities in terms of focusing the radiation on the target volume and sparing surrounding healthy structures [14,15].

Especially for BP-RMS, two different BT treatment combinations have been published in larger numbers of cases, which vary predominantly with regard to the extent of surgery and the irradiation technique used [8,16]. In our department, we aim for a marginal R0 resection and use individually 3-dimensional (3D)-planned high-dose rate (HDR)-BT. A major challenge, besides the actual organ-preserving tumor resection and the subsequent reconstruction of the organ by the surgeon, is the exact placement of the BT tubes around the former tumor area, without injury to the adjacent healthy organs, in typically young children with correspondingly small anatomical conditions. Incorrect positioning of the tubes would lead to inadequate irradiation of the target volume as well as unnecessary irradiation of the surrounding healthy structures and therefore must be avoided. The exact position of the tubes can only be determined with certainty by means of an MRI or CT scan. Up until now, we have carried out an MRI outside the operation room (OR) unit. If a tube revision was necessary, the patient had to be brought back to the OR. In order to improve the patient safety and the treatment quality, we evaluated the use of intraoperative magnetic resonance imaging (iMRI) for controlling the position of the surgically inserted BT tubes and planning the irradiation. To our knowledge, this is the first report on the use of iMRI in children with BP and perianal RMS.

iMRI was initially developed for neurosurgical procedures and is frequently used in combination with neuronavigation. Indications for its use in children are gliomas, pituitary adenoma, vascular diseases, and epilepsy surgery [17–19]. In most cases, it serves to control the extent of resection and correct anatomical changes after tumor resection due to the so-called brain shift [20,21].

## 2. Materials and Methods

All patients with BP and perianal RMS who qualified for organ-preserving tumor resection and BT, and were treated in our department since January 2021 were prospectively registered in this study. A historical cohort consisted of all patients who received analogous treatment from 2009 to December 2020, in which the MRI for controlling the position of the BT tubes was performed outside the OR unit [10]. Patients who received only a CT scan to control the BT tube position and for BT planning are not included in the study (10 BP-RMS, 3 perianal RMS). The data analysis was done retrospectively. The study was approved by the institutional ethics committee (No. 293/2023BO2).

All patients received risk-adjusted neoadjuvant chemotherapy according to the current Cooperative Weichteilsarkom Studiengruppe (CWS) protocol [22].

The local therapy was determined in a multidisciplinary tumor board (MDT). It was primarily based on the preoperatively evaluated tumor extension and the intraoperative findings. Patients were included if organ-preserving tumor resection with subsequent HDR-BT was performed.

For irradiation in BP-RMS patients, BT tubes are placed after tumor resection around the former tumor area, with an additional tube being inserted transurethrally in individual cases (Figure 1). The tube placement is performed in close cooperation and coordination between the surgeon and the treating radiation oncologists. A sharp cannula is inserted into the small pelvis through the perineum (Figure 1A–D). The BT tube is then placed through this cannula and fixed to the outside of the skin as well as inside the body with absorbable sutures (Figure 1E,F). Children under 3 years of age pose a particular challenge due to the anatomic conditions of a small caliber urethra and a narrow retrourethral spatium, with an increased risk of perforation of the rectum and injury to the corpora cavernosa. After tumor resection in children with perianal RMS, we place the tubes at the resection margin and fix them there. With the individually 3D-planned HDR-BT technique used, the distance between the tubes should not exceed 1 cm, as otherwise an excessively high punctual irradiation dose is necessary to adequately cover the previously determined target volume. BT should typically start on the second day after surgery. In fractions of 3 Gy, administered twice daily, the children received a total dose of 36 Gy. Details of the protocol have been described before [8].

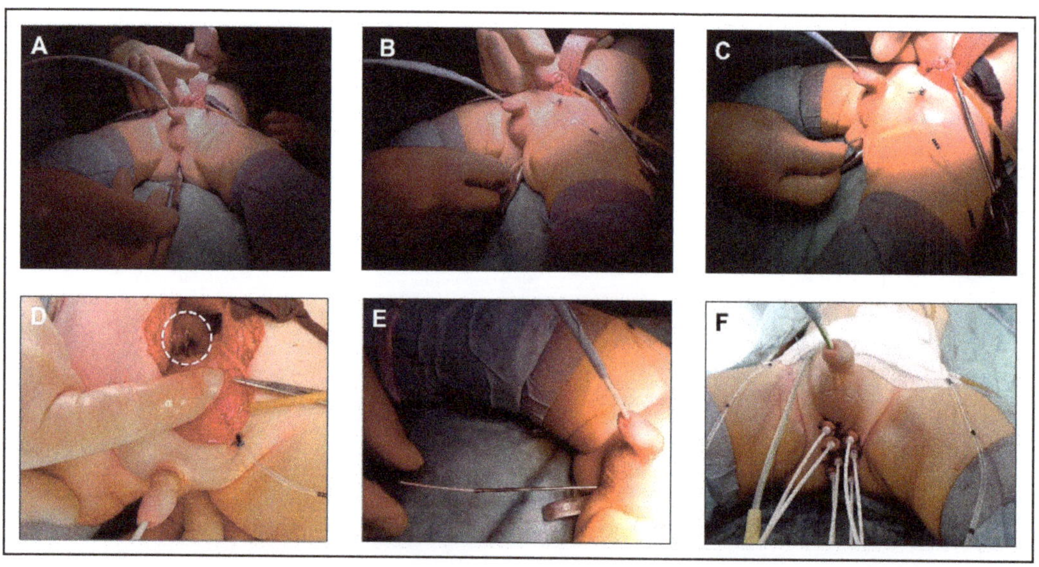


**Figure 1.** Principle of brachytherapy tube placement. (**A–C**) insertion of a sharp cannula into the small pelvis through the perineum. For identification, a metal probe was inserted into the rectum. (**D**) tip of the cannula (dashed line) in the small pelvis behind the urinary bladder. (**E**) insertion of the BT tube through the cannula. (**F**) perineum after placement of 7 BT tubes and fixation to the skin. An additional tube was inserted transurethrally.

To control the position of the BT tubes and to plan the irradiation, an MRI and a CT scan are performed after the placement of the tubes. The additional CT scan is necessary for the exact contouring of the tubes and the BT planning as well as radiation dose calculation.

A modified ceiling-mounted, moveable 1.5 T magnet (Espree; Siemens Medical Systems, Erlangen, Germany) was used for iMRI (Figure 2). It is located in an intraoperative

MR suite (IMRIS Visius Surgical Theatre; IMRIS Inc., Winnipeg, MB, Canada) [23]. The magnet is located in a "parking bay" with shielded doors. In this condition and with the doors closed, the adjacent OR can be used in a regular manner. Once the magnet moves to its "scanning position", all ferromagnetic equipment must be placed outside the 5 Gauss line. For neurosurgical procedures, the patient's head is placed at the top of the table in an MR-compatible DORO skull clamp with disposable skull pins (ProMed Instruments GmbH, Freiburg, Germany). Since the "scanning position" of the magnet is predetermined and cannot be changed, we placed the children in our study the other way around on the operating table, meaning with their feet pointing towards the MRI (Figure 2A,B). To prevent the feet from hanging in the air, a non-ferromagnetic, MR-compatible table extension was designed specifically for this purpose. All monitoring equipment was MR-compatible, and hearing protection was achieved by using ear plugs (Ohropax Yellow, OHROPAX GmbH, Wehrheim, Germany) combined with neonatal noise attenuators (MiniMuffs, Natus Medical Incorporated, Oakville, ON, Canada). Adequate preparation of the patient and the OR was checked with standardized MR safety checklists, before moving the patient into the scanning position.

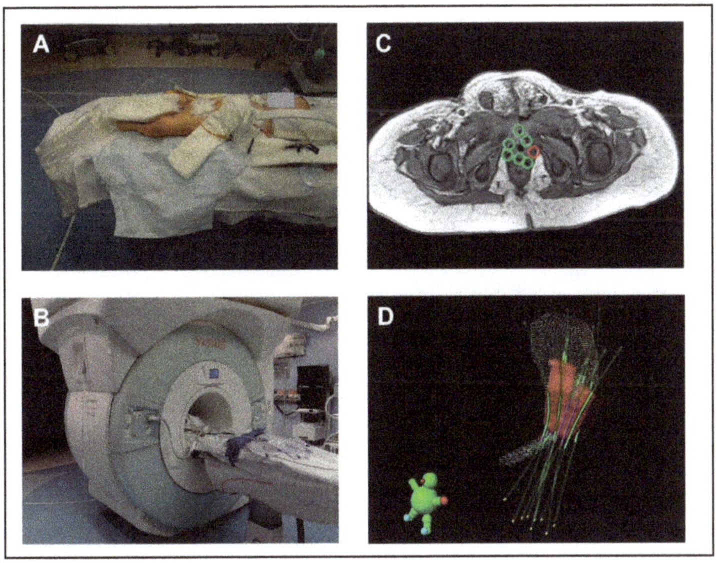


**Figure 2.** Principle of intraoperative MRI. (**A**) iMRI suite; position of a child on the table extension designed for this specific purpose. (**B**) "scanning position" of the magnet. (**C**) iMR image with BT tubes. The green marked tubes are in the correct position; a tube is missing at the red marked site. (**D**) 3D image of the BT tube location in relation to the surrounding anatomical structures in the radiotherapy planning system. Red volume = target volume, blue volume = urethra, yellow volume = urinary bladder, brown volume = rectum, green lines = BT tubes.

The MRI protocol included, in all patients, a transversal T2-weighted turbospin-echo (TSE) sequence (49 slices, 3 mm slice thickness, repetition time (TR) 10,860 ms, echo time (TE) 127 ms, field-of-view (FoV) 180 $\times$ 153 mm$^2$, in-plane resolution 0.56 $\times$ 0.56 mm$^2$), a sagittal T2-weighted TSE (29 slices, 3 mm slice thickness, TR 5500 ms, TE 129 ms, FoV 180 $\times$ 151 mm$^2$, in-plane resolution 0.47 $\times$ 0.47 mm$^2$), and a T1-weighted transversal TSE sequence (49 slices, 3 mm slice thickness, TR 661 ms, TE 15 ms, FoV 160 $\times$ 148 mm, in-plane resolution 0.5 $\times$ 0.5 mm$^2$) that cover the complete pelvic region. Depending on preoperative imaging, additional diffusion-weighted images, T1 fat saturated images or T2-weighted fat saturated images, were acquired on a case-by-case basis.

## 3. Results

### 3.1. Patients

Over a period of 25 months, 12 children (12 boys) were included in this study. Median age at time of surgery and BT was 28 months (range 13–73). Ten of the patients had a BP-RMS and two patients had a perianal RMS. All tumors were of the embryonal subtype. After a median follow-up of 10.5 months (range 6–23), 10 patients are still alive and show no evidence of tumor relapse (Table 1). One of the patients with BP-RMS experienced an early relapse and died despite maximum therapy due to tumor progress. A second patient experienced an intestinal obstruction with a septic shock seven months after tumor resection and died. One patient with BP-RMS developed urethral stenosis in the postoperative course. He has since been successfully operated on and can void without any problems. No other postoperative complications have been observed.

**Table 1.** Patient characteristics and outcome.

| Patient | Diagnosis | Age at Local Therapy (Months) | Weight at Local Therapy (kg) | Follow-Up (Months) | Follow-Up Findings |
|---|---|---|---|---|---|
| 1 | BP-RMS | 21 | 10.3 | 23 | NED |
| 2 | Perianal RMS | 20 | 11.9 | 22 | NED |
| 3 | BP-RMS | 28 | 11.4 | 7 | Dead [1] |
| 4 | BP-RMS | 35 | 10.5 | 8 | Dead [2] |
| 5 | BP-RMS | 28 | 12.0 | 16 | NED |
| 6 | BP-RMS | 32 | 12.6 | 7 | NED |
| 7 | BP-RMS | 24 | 10.8 | 13 | NED |
| 8 | BP-RMS | 57 | 18.5 | 12 | NED |
| 9 | BP-RMS | 30 | 12.0 | 9 | NED |
| 10 | BP-RMS | 13 | 9.0 | 6 | NED |
| 11 | Perianal RMS | 22 | 14.6 | 0 | N/A |
| 12 | BP-RMS | 73 | 20.0 | 0 | N/A |

BP bladder/prostate; RMS rhabdomyosarcoma; NED no evidence of disease; N/A not available due to follow-up period being too short; [1] due to tumor progress; [2] non oncological reason.

### 3.2. iMRI

The time from the end of surgery to the start of the iMRI examination was approximately 30 min. During this time, the child was prepared for the iMRI and all necessary arrangements were made for the iMRI to move into the "scanning position". The total iMRI scan time was about 14 min for the three basic sequences, with a maximum duration of around 25 min with additional sequences. In all patients, the important anatomical structures (urinary bladder, prostate, rectum, urethra, ureters, and intestine) could be visualized in adequate quality. The iMRI allowed precise anatomical localization of the BT tubes and an accurate contouring of the target volume. No patient experienced an adverse event during the scan period.

### 3.3. BT Tubes

A median of eight tubes were placed for BT (range 7–10). In 6 out of 12 patients, BT tubes had to be revised. All tube misplacements were detected by iMRI. The attending radiologist and the radiation oncologist in charge indicated the need for revision. The reasons for this were too large of a distance between two tubes in five cases and a tube placed in the rectal wall in one case. In none of these six patients did a delay in the start of irradiation occur. In the historical group with a total of 20 patients (20 BP-RMS), correction or new placement of BT tubes was necessary in 5 patients. In these cases, the main reason for revision was also a distance between two tubes that was too large (n = 5). In three of these patients, the irradiation start was delayed and could not be started until postoperative day 3.

## 4. Discussion

This study describes our initial experience with an iMRI for BT tube control and irradiation planning in children with BP and perianal RMS. So far, we have successfully applied this concept in 12 children with these diagnoses.

The difficulty that arises when placing the BT tubes, especially in children with BP-RMS, is the fact that the course of the tubes cannot be seen over a certain distance. In most cases, however, this is exactly the critical area and thereby a large part of the target volume of the irradiation field, in which exact positioning of the tubes is essential. Since we have been using the concept of organ-preserving tumor resection in combination with 3D-planned BT in our department, cross-sectional imaging has been used to control the position of the tubes and to plan the irradiation. Initially, we carried out only a CT scan and in the course of time we added an MRI. The reason for this was the superior tissue resolution of MRI compared to CT. Anatomical structures, especially the organs at risk such as the rectum, urethra, etc., and the position of the BT tubes in relation to the anatomical structures can be better visualized. In addition, the target volume for irradiation can be more precisely defined. The decision of whether a tube revision is necessary can be made with MRI alone. Since January 2021, we have been using iMRI. For technical reasons, we cannot yet dispense with the CT as it is needed for exact radiation dose calculations based on electron densities of different tissues and the exact position of the radioactive sources. An adaptation of the BT planning software and spatial resolution of MRI might make this possible in the future. The main advantages of not using CT would be additional time saving and less radiation exposure for the young patients. Up until now, the two methods have been considered complementary.

Compared to HDR-BT, the pulsed dose rate (PDR)-BT used in Paris seems to have an advantage in terms of BT tube placement—fewer tubes are used for irradiation and are being placed with more standardization and with less defined distances [11]. The differences in the influence of precise positioning with advantage for PDR might be explained by the lower single dose application per fraction and the estimated repair capacity between the pulses. For PDR, certain structural and personnel requirements are necessary, which are not given everywhere, including at our institution. However, with regard to the oncological and also the functional outcome, there are no relevant differences when using the two different concepts of organ-preserving tumor resection and BT [24,25].

By using iMRI and thus eliminating the need for transportation between the OR unit and the MRI, we were able to save a lot of time and start radiation therapy in all patients at the scheduled time. In the historical cohort, this was not possible in 3 out of 20 patients, mainly because of the use and availability of the MRI outside the OR unit and the fact that potential difficult revisions of BT tubes had to be performed the day after when the experienced surgical team was available again. It cannot be assumed that a delayed start of irradiation by one day leads to a worse oncological outcome. However, the fact that our patients have to be sedated and undergo muscle relaxation from the beginning of the tumor resection until the end of BT in order to prevent BT tube dislocation, prolongs the time under anesthesia and likely increases the associated side effects [26].

iMRI cannot prevent misplacement of BT tubes but gives the ability to correct the misplacement within the same procedure. We have investigated different methods to optimize tube placement, like transrectal sonography, and are still evaluating different other modalities. However, we have not yet found a suitable tool that displays the anatomical structures as precisely as an MRI.

In the literature, there are other examples of iMRI application in children, such as in children with imperforate anus [27–29], and bladder exstrophy [30]. However, this imaging technique has not become generally accepted for these indications. In our study, iMRI has been proven to be effective in tube control and BT planning in children with BP and perianal RMS. It is now used as a standard for all children with these diagnoses and after the combination treatment of organ-preserving tumor resection and BT. In the future, we are planning to expand the use of iMRI for other indications as well.

This study is the first report of the use of iMRI in the combination therapy of BT and organ-preserving tumor resection in children with BP and perianal RMS and describes the experience of a single center with this method. This study has limitations. The results are based on a small number of cases due to the low incidence of the diseases studied and the even lower number of patients who are suitable for this individualized combination therapy [1,2,9]. The incidence is even less in perianal RMS than in BP-RMS [9]. A separate report on perianal RMS would not be meaningful, so these patients were included in this study. This also seems justified as the focus of the study is on the use of iMRI as a tool to improve the clinical workflow of the combination therapy, which is performed in a similar way in both diseases.

Due to the small number of cases and the low statistical power, the analysis was descriptive only. Patients were prospectively enrolled in the study. However, it was not possible to randomize a significant number of patients, so historical data were used for comparison. This study must therefore be regarded as the first report of the use of a new method for a known application, as a proof of principle.

## 5. Conclusions

In this initial study, by using iMRI, we have been able to increase patient safety and therapy quality by eliminating risky transports under anesthesia between OR unit and the MRI. In addition, delay of radiotherapy start and thus prolonged time under anesthesia can be avoided with this approach in case of necessary surgical revision.

**Author Contributions:** Conceptualization, A.S., C.R., J.F., F.P. and B.B.; validation, F.E., D.B., M.N., F.F., U.E., M.T., J.S., C.U., S.S. and J.F.; data curation, A.S., C.R., C.U. and B.B.; writing—original draft preparation, A.S., C.R., F.P. and B.B.; writing—review and editing, F.E., D.B., M.N., F.F., U.E., M.T., J.S., C.U., S.S. and J.F.; visualization, A.S., S.S. and F.P.; supervision, U.E., M.T., M.N. and J.F.; project administration, A.S. All authors have read and agreed to the published version of the manuscript.

**Funding:** This research received no external funding.

**Institutional Review Board Statement:** The study was conducted in accordance with the Declaration of Helsinki, and approved by the Ethics Committee at the Faculty of Medicine of the Eberhard-Karls University and the University Hospital Tuebingen (No. 293/2023BO2).

**Informed Consent Statement:** Informed consent was obtained from all subjects involved in the study.

**Data Availability Statement:** The data presented in this study are available in this article.

**Conflicts of Interest:** The authors declare no conflict of interest.

## References

1. Dasgupta, R.; Fuchs, J.; Rodeberg, D. Rhabdomyosarcoma. *Semin. Pediatr. Surg.* **2016**, *25*, 276–283. [CrossRef]
2. Arndt, C.; Rodeberg, D.; Breitfeld, P.P.; Raney, R.B.; Ullrich, F.; Donaldson, S. Does bladder preservation (as a surgical principle) lead to retaining bladder function in bladder/prostate rhabdomyosarcoma? Results from intergroup rhabdomyosarcoma study iv. *J. Urol.* **2004**, *171*, 2396–2403. [CrossRef]
3. Terwisscha van Scheltinga, S.; Rogers, T.; Smeulders, N.; deCorti, F.; Guerin, F.; Craigie, R.; Burrieza, G.G.; Smeele, L.; Hol, M.; van Rijn, R.; et al. Developments in the Surgical Approach to Staging and Resection of Rhabdomyosarcoma. *Cancers* **2023**, *15*, 449. [CrossRef]
4. Ferrer, F.A.; Isakoff, M.; Koyle, M.A. Bladder/prostate rhabdomyosarcoma: Past, present and future. *J. Urol.* **2006**, *176*, 1283–1291. [CrossRef]
5. Shapiro, D.D.; Harel, M.; Ferrer, F.; McKenna, P.H. Focusing on organ preservation and function: Paradigm shifts in the treatment of pediatric genitourinary rhabdomyosarcoma. *Int. Urol. Nephrol.* **2016**, *48*, 1009–1013. [CrossRef]
6. Alexander, N.; Lane, S.; Hitchcock, R. What is the evidence for radical surgery in the management of localized embryonal bladder/prostate rhabdomyosarcoma? *Pediatr. Blood Cancer* **2012**, *58*, 833–835. [CrossRef]

7. Magne, N.; Oberlin, O.; Martelli, H.; Gerbaulet, A.; Chassagne, D.; Haie-Meder, C. Vulval and vaginal rhabdomyosarcoma in children: Update and reappraisal of Institut Gustave Roussy brachytherapy experience. *Int. J. Radiat. Oncol. Biol. Phys.* **2008**, *72*, 878–883. [CrossRef]
8. Fuchs, J.; Paulsen, F.; Bleif, M.; Lamprecht, U.; Weidner, N.; Zips, D.; Neunhoeffer, F.; Seitz, G. Conservative surgery with combined high dose rate brachytherapy for patients suffering from genitourinary and perianal rhabdomyosarcoma. *Radiother. Oncol.* **2016**, *121*, 262–267. [CrossRef]
9. Rogers, T.; Zanetti, I.; Coppadoro, B.; Martelli, H.; Jenney, M.; Minard-Colin, V.; Terwisscha van Scheltinga, S.E.J.; Skerritt, C.; Fajardo, R.D.; Guerin, F.; et al. Perianal/perineal rhabdomyosarcoma: Results of the SIOP MMT 95, Italian RMS 96, and EpSSG RMS 2005 studies. *Pediatr. Blood Cancer* **2022**, *69*, e29739. [CrossRef]
10. Schmidt, A.; Warmann, S.W.; Eckert, F.; Ellerkamp, V.; Schaefer, J.; Blumenstock, G.; Paulsen, F.; Fuchs, J. The Role of Reconstructive Surgery and Brachytherapy in Pediatric Bladder/Prostate Rhabdomyosarcoma. *J. Urol.* **2020**, *204*, 825–834. [CrossRef]
11. Chargari, C.; Haie-Meder, C.; Guerin, F.; Minard-Colin, V.; de Lambert, G.; Mazeron, R.; Escande, A.; Marsolat, F.; Dumas, I.; Deutsch, E.; et al. Brachytherapy Combined With Surgery for Conservative Treatment of Children With Bladder Neck and/or Prostate Rhabdomyosarcoma. *Int. J. Radiat. Oncol. Biol. Phys.* **2017**, *98*, 352–359. [CrossRef] [PubMed]
12. Raney, R.B., Jr.; Donaldson, M.H.; Sutow, W.W.; Lindberg, R.D.; Maurer, H.M.; Tefft, M. Special considerations related to primary site in rhabdomyosarcoma: Experience of the Intergroup Rhabdomyosarcoma Study, 1972--76. *Natl. Cancer Inst. Monogr.* **1981**, *56*, 69–74.
13. Fuchs, J.; Dantonello, T.M.; Blumenstock, G.; Kosztyla, D.; Klingebiel, T.; Leuschner, I.; Schuck, A.; Niggli, F.K.; Koscielniak, E.; Seitz, G. Treatment and outcome of patients suffering from perineal/perianal rhabdomyosarcoma: Results from the CWS trials--retrospective clinical study. *Ann. Surg.* **2014**, *259*, 1166–1172. [CrossRef]
14. Heinzelmann, F.; Thorwarth, D.; Lamprecht, U.; Kaulich, T.W.; Fuchs, J.; Seitz, G.; Ebinger, M.; Handgretinger, R.; Bamberg, M.; Weinmann, M. Comparison of different adjuvant radiotherapy approaches in childhood bladder/prostate rhabdomyosarcoma treated with conservative surgery. *Strahlenther. Onkol.* **2011**, *187*, 715–721. [CrossRef] [PubMed]
15. Chargari, C.; Deutsch, E.; Blanchard, P.; Gouy, S.; Martelli, H.; Guerin, F.; Dumas, I.; Bossi, A.; Morice, P.; Viswanathan, A.N.; et al. Brachytherapy: An overview for clinicians. *CA Cancer J. Clin.* **2019**, *69*, 386–401. [CrossRef]
16. Chargari, C.; Martelli, H.; Guerin, F.; Bacorro, W.; de Lambert, G.; Escande, A.; Minard-Colin, V.; Dumas, I.; Deutsch, E.; Haie-Meder, C. Pulsed-dose rate brachytherapy for pediatric bladder prostate rhabdomyosarcoma: Compliance and early clinical results. *Radiother. Oncol.* **2017**, *124*, 285–290. [CrossRef]
17. Levy, R.; Cox, R.G.; Hader, W.J.; Myles, T.; Sutherland, G.R.; Hamilton, M.G. Application of intraoperative high-field magnetic resonance imaging in pediatric neurosurgery. *J. Neurosurg. Pediatr.* **2009**, *4*, 467–474. [CrossRef] [PubMed]
18. Yousaf, J.; Avula, S.; Abernethy, L.J.; Mallucci, C.L. Importance of intraoperative magnetic resonance imaging for pediatric brain tumor surgery. *Surg. Neurol. Int.* **2012**, *3*, S65–S72. [CrossRef]
19. Gohla, G.; Bender, B.; Tatagiba, M.; Honegger, J.; Ernemann, U.; Roder, C. Identification of tumor residuals in pituitary adenoma surgery with intraoperative MRI: Do we need gadolinium? *Neurosurg. Rev.* **2020**, *43*, 1623–1629. [CrossRef]
20. Roder, C.; Bisdas, S.; Ebner, F.H.; Honegger, J.; Naegele, T.; Ernemann, U.; Tatagiba, M. Maximizing the extent of resection and survival benefit of patients in glioblastoma surgery: High-field iMRI versus conventional and 5-ALA-assisted surgery. *Eur. J. Surg. Oncol.* **2014**, *40*, 297–304. [CrossRef] [PubMed]
21. Roder, C.; Breitkopf, M.; Bisdas, S.; Freitas Rda, S.; Dimostheni, A.; Ebinger, M.; Wolff, M.; Tatagiba, M.; Schuhmann, M.U. Beneficial impact of high-field intraoperative magnetic resonance imaging on the efficacy of pediatric low-grade glioma surgery. *Neurosurg. Focus* **2016**, *40*, E13. [CrossRef] [PubMed]
22. Cooperative Weichteilsarkom Studiengruppe (CWS) of the Gesellschaft für Pädiatrische Onkologie und Hämatologie (GPOH). *CWS-Guidance for Risk Adapted Treatment of Soft Tissue Sarcoma and Soft Tissue Tumours in Children, Adolescents, and Young Adults*; Cooperative Weichteilsarkom Studiengruppe (CWS) of the Gesellschaft für Pädiatrische Onkologie und Hämatologie (GPOH): Stuttgart, Germany, 2014.
23. Chen, X.; Xu, B.N.; Meng, X.; Zhang, J.; Yu, X.; Zhou, D. Dual-room 1.5-T intraoperative magnetic resonance imaging suite with a movable magnet: Implementation and preliminary experience. *Neurosurg. Rev.* **2012**, *35*, 95–109; discussion 109–110. [CrossRef] [PubMed]
24. Akkary, R.; Guerin, F.; Chargari, C.; Jochault, L.; Audry, G.; Pio, L.; Minard-Colin, V.; Haie-Meder, C.; Martelli, H. Long-term urological complications after conservative local treatment (surgery and brachytherapy) in children with bladder-prostate rhabdomyosarcoma: A single-team experience. *Pediatr. Blood Cancer* **2022**, *69*, e29532. [CrossRef]
25. Ellerkamp, V.; Schmidt, A.; Warmann, S.W.; Eckert, F.; Schaefer, J.; Paulsen, F.; Fuchs, J. Detailed functional results after bladder-preserving surgery and high-dose-rate brachytherapy in pediatric bladder/prostate rhabdomyosarcoma. *J. Cancer Res. Clin. Oncol.* **2022**, *149*, 3161–3170. [CrossRef] [PubMed]
26. Michel, J.; Sauter, L.; Neunhoeffer, F.; Hofbeck, M.; Kumpf, M.; Paulsen, F.; Schmidt, A.; Fuchs, J. Sedation practices during high dose rate brachytherapy for children with urogenital and perianal rhabdomyosarcoma. *J. Pediatr. Surg.* **2022**, *57*, 1432–1438. [CrossRef] [PubMed]
27. Thomas, T.T.; Teitelbaum, D.H.; Smith, E.A.; Dillman, J.R.; Vellody, R.; Jarboe, M.D. Magnetic resonance imaging (MRI)-assisted laparoscopic anorectoplasty for imperforate anus: A single center experience. *Pediatr. Surg. Int.* **2017**, *33*, 15–21. [CrossRef]

28. Raschbaum, G.R.; Bleacher, J.C.; Grattan-Smith, J.D.; Jones, R.A. Magnetic resonance imaging-guided laparoscopic-assisted anorectoplasty for imperforate anus. *J. Pediatr. Surg.* **2010**, *45*, 220–223. [CrossRef] [PubMed]
29. Jarboe, M.; Ladino-Torres, M.; Wild, L.; Spremo, D.; Elkins, S.; Ladouceur, R.; Nagy, D.; Ehrlich, P.; Ralls, M. Imaged-guided and muscle sparing laparoscopic anorectoplasty using real-time magnetic resonance imaging. *Pediatr. Surg. Int.* **2020**, *36*, 1255–1260. [CrossRef]
30. Di Carlo, H.N.; Maruf, M.; Massanyi, E.Z.; Shah, B.; Tekes, A.; Gearhart, J.P. 3-Dimensional Magnetic Resonance Imaging Guided Pelvic Floor Dissection for Bladder Exstrophy: A Single Arm Trial. *J. Urol.* **2019**, *202*, 406–412. [CrossRef]

cancers

MDPI

Article

# Prognostic Factors of Platinum-Refractory Advanced Urothelial Carcinoma Treated with Pembrolizumab

Yasunori Akashi [1], Yutaka Yamamoto [1,*], Mamoru Hashimoto [2], Shogo Adomi [2], Kazutoshi Fujita [2], Keisuke Kiba [1], Takafumi Minami [2], Kazuhiro Yoshimura [2], Akihide Hirayama [1] and Hirotsugu Uemura [2]

1 Department of Urology, Kindai University Nara Hospital, Ikoma 630-0293, Japan; 144053@med.kindai.ac.jp (Y.A.)
2 Department of Urology, Kindai University Hospital, Osakasayama 589-8511, Japan
* Correspondence: yamamotokindai@yahoo.co.jp; Tel.: +81-74-377-0880

**Simple Summary:** Several studies have investigated various types of biomarkers to predict responses to immune checkpoint inhibitor (ICI) therapy for patients with platinum-refractory advanced urothelial carcinoma, but they were inconclusive. Recently, antibiotic exposure has attracted attention as a biomarker because it may affect antitumor immunity through changes in gut microbiota. We evaluated the factors predictive of ICI response, including antibiotic exposure, in 41 metastatic urothelial carcinoma patients. The patients' median age was 75 years, and the vast majority of the patients were male. The objective response rate was 29.3%, with a median overall survival (OS) of 17.8 months. A high neutrophil-to-lymphocyte ratio (NLR) and poor performance status (PS) were significantly associated with poor OS. Antibiotic exposure did not have a significant impact on OS.

**Abstract:** Introduction: Immune checkpoint inhibitor (ICI) therapy has significantly improved the prognosis of some patients with advanced urothelial carcinoma (UC), but it does not provide high therapeutic efficacy in all patients. Therefore, identifying predictive biomarkers is crucial in determining which patients are candidates for ICI treatment. This study aimed to identify the predictors of ICI treatment response in patients with platinum-refractory advanced UC treated with pembrolizumab. Methods: Patients with platinum-refractory advanced UC who had received pembrolizumab at two hospitals in Japan were included. Univariate and multivariate analyses were performed to identify biomarkers for progression-free survival (PFS) and overall survival (OS). Results: Forty-one patients were evaluable for this analysis. Their median age was 75 years, and the vast majority of the patients were male (85.4%). The objective response rate was 29.3%, with a median overall survival (OS) of 17.8 months. On multivariate analysis, an Eastern Cooperative Oncology Group performance status (ECOG-PS) $\geq 2$ (HR = 6.33, $p = 0.03$) and a baseline neutrophil-to-lymphocyte ratio (NLR) > 3 (HR = 2.79, $p = 0.04$) were significantly associated with poor OS. Antibiotic exposure did not have a significant impact on either PFS or OS. Conclusions: ECOG-PS $\geq 2$ and baseline NLR > 3 were independent risk factors for OS in patients with platinum-refractory advanced UC treated with pembrolizumab. Antibiotic exposure was not a predictor of ICI treatment response.

**Keywords:** advanced urothelial cancer; immune checkpoint inhibitor; antibiotic exposure

## 1. Introduction

Gemcitabine, cisplatin (GC) and methotrexate, vinblastine, doxorubicin, and cisplatin (MVAC) are widely used as first-line regimens in metastatic urothelial cancer (UC), but many cases are refractory [1]. After the failure of platinum-based chemotherapy, there was no internationally accepted standard of care. However, since the advent of an immune checkpoint inhibitor (ICI) for advanced UC, the treatment strategy for UC patients has changed dramatically. Pembrolizumab is a humanized monoclonal IgG4κ isotype antibody that directly inhibits programmed cell death-1 (PD-1) and its ligands, programmed cell death ligand-1 (PD-L1) and programmed cell death ligand-2 (PD-L2). PD-1 inhibits cytokine production from T cells and cell proliferation by binding to PD-L1, which results in the suppression of immune responses [2]. PD-L1 also suppresses immune responses by converting naïve CD4(+) T cells into regulatory T (Treg) cells [3]. Pembrolizumab inhibits the binding of PD-1 to both PD-L1 and PD-L2 ligands, activating cancer-specific cytotoxic T lymphocytes and the PD-L1 and PD-L2 ligands.

In the randomized phase 3 Keynote-045 trial, pembrolizumab showed better outcomes than chemotherapy as a second-line therapy for UC patients who progressed after platinum-containing regimens [4]. Based on these results, current guidelines recommend pembrolizumab as a second-line systemic therapy after platinum-containing regimens [5]. However, pembrolizumab offers an objective response rate (ORR) of approximately 20%, which is anything but satisfactory. Identification of predictive biomarkers could increase the benefit of ICI treatment and avoid therapeutic intervention if the likelihood of response is predicted to be low. Therefore, it is crucial to identify biomarkers that predict which patients will benefit from pembrolizumab treatment. Promising biomarkers for predicting the response to ICI therapy include the expression of programmed death ligand-1 (PD-L1), the tumor mutational burden (TMB), and circulating tumor DNA in various types of cancer [6–8]. However, an ideal biomarker would be reproducible, cost-effective, and simple. From this perspective, the usefulness of the Eastern Cooperative Oncology Group performance status (ECOG-PS), location of the metastasis, C-reactive protein (CRP), and the neutrophil-to-lymphocyte ratio (NLR) has been studied in several malignancies, including UC [9–16]. More recently, one factor that is attracting interest is the association between ICI treatment outcomes and antibiotic exposure. This is due to the fact that antibiotic exposure alters the gut microbiome, which in turn affects the effectiveness and immune-related toxicities of ICIs [17–22]. However, few reports have evaluated the association between antibiotic exposure and ICI treatment outcomes of patients with UC [23]. The purpose of this study was to investigate potential predictive biomarkers, including antibiotic exposure, in patients with UC treated with pembrolizumab, which could provide useful information for patients who require ICI treatment.

## 2. Materials and Methods

### 2.1. Patients and Clinical Data

The study cohort consisted of advanced or metastatic UC patients who received pembrolizumab at Kindai University Nara Hospital and Kindai University Hospital in Japan between January 2018 and December 2021. All patients received a pathological diagnosis of UC and were treated with platinum-containing neoadjuvant or adjuvant chemotherapy. All samples enrolled in this study were acquired via surgical resection or biopsy. Clinicopathological data were obtained from the patients' medical records. Pembrolizumab 200 mg was administered intravenously every three weeks, and it was continued until unacceptable toxicity or either radiographic or clinical disease progression. This study was approved by the Institutional Review Board of Kindai University Nara Hospital and Kindai University Hospital (approval number 705). Informed consent was waived because of the retrospective design of the study. However, an opt-out opportunity for this study was provided through the website of our institution (https://www.med.kindai.ac.jp/uro/, accessed on 5 December 2023).

### 2.2. Statistical Analysis

Progression-free survival (PFS) and overall survival (OS) were calculated with the Kaplan–Meier method. The ORR to pembrolizumab was assessed according to the response evaluation criteria in solid tumors (RECIST) version 1.1 [24]. Adverse events (AEs) following pembrolizumab were evaluated at each visit during and after treatment. The severity of the AEs was graded according to the National Cancer Institute's common terminology criteria for adverse events (CTCAE) v5.0 [25]. Univariate and multivariate analysis were performed using a Cox proportional hazards model to identify the biomarkers for PFS and OS. A $p$-value < 0.05 was considered significant. The Statistical Package for the Social Sciences version 23.0 (SPSSs, Chicago, IL, USA) was used for all statistical analysis.

## 3. Results

### 3.1. Patients' Characteristics

A total of 41 patients met the eligibility criteria and were evaluable for this analysis. The median duration of pembrolizumab treatment was 4.0 (0–25.0 PS) months, and the median follow-up period was 16.5 (range, 1.0–47.8) months. No patients received atezolizumab as a second-line therapy after chemotherapy because it was not covered by the Japanese health insurance system. Table 1 shows the patients' baseline characteristics. The patients' median age was 75 (range, 58–81) years, and the vast majority of the patients were male (85.4%). Sixteen of the forty-one patients had upper UC. Twenty-nine patients had undergone total cystectomy or total nephroureterectomy, and eight patients had received radiation therapy. ECOG-PS was 0 or 1 in 90.2% (37/41) of patients. The median number of cycles of pembrolizumab treatment was five (range, 1–32). Lymph nodes were the most common sites of metastases (66%, 27/41), followed by the lungs (49%, 20/41), bones (22%, 9/41), and liver (15%, 6/41). The median neutrophil-to-lymphocyte ratio (NLR) at baseline was 2.96 (range, 1.27–28.4). Antibiotic exposure was defined as antibiotic use for at least 1 week within 1 month before or after starting pembrolizumab. Sixteen patients (39%) were identified as the antibiotic exposure cohort, with a median duration of antibiotic exposure of 7 (range, 7–30) days. With regard to antibiotic classes, they included cephalosporins in seven cases, fluoroquinolones in five cases, and penicillins in four cases. The most common reasons for the use of antibiotics were urinary tract infections (24%), followed by other infections (12.2%) or febrile neutropenia (2.4%).

**Table 1.** Characteristics of patients included in the analysis.

| Variable | Patients ($n$ = 41) |
|---|---|
| Age (years), median (range) | 75 (58~81) |
| Observation period (months), median (range) | 16.5 (1.0~47.8) |
| Male sex, $n$ (%) | 35 (85.4%) |
| Site of primary tumor, Upper urinary tract/Bladder, $n$ (%) | 16/25 (40%/60%) |
| Response criteria (RECIST), CR/PR/SD/PD, $n$ (%) | 7/5/2/29 (17%/12%/5%/70%) |
| ECOG-PS, 0/1/2, $n$ (%) | 27/10/4 (66%/24%/10%) |
| Number of prior regimens, 1/2/3/4, $n$ (%) | 20/14/6/1 (82%/8%/8%/2%) |
| Metastatic sites, liver/lung/bone/lymph node, $n$ (%) | 6/20/9/27 (15%/49%/22%/66%) |
| Number of metastatic organs, 1/2/3/4, $n$ (%) | 20/14/6/1 (48%/34%/15%/2%) |
| Hemoglobin, > 10 mg/dL/ < 10 ng/dL, $n$ (%) | 33/8 (80%/20%) |
| CRP baseline (mg/dL), median (range) | 0.56 (0.03~21) |
| NLR baseline, median (range) | 2.96 (1.27~28.4) |
| Antibiotic exposure, $n$ (%) | 16 (39%) |
| Duration of antibiotic exposure (days), median (range) | 7 (7–30) |
| Antibiotic classes, Cephalosporin/fluoroquinolone/penicillin, $n$ (%) | 7/5/4 (44%/31%/25%) |

### 3.2. *Efficacy of Pembrolizumab and Adverse Events*

The OS and PFS are shown in Figure 1A,B. Thirty-one patients (75.6%) discontinued treatment with pembrolizumab because of disease progression (70.7%) and immune-related AEs (4.9%). Twenty-five patients (60.9%) had died at the time of analysis. The median PFS and OS were 4.9 [95% confidence interval (CI), 1.2–8.6] months and 17.8 [95% CI, 11.5–24.0] months, respectively. The causes of death of the 25 patients who died were UC in 24 (96%)

and suicide in 1 patient (4%). The best overall response assessed according to RECIST is shown in Table 2. Totals of 7 (17.1%), 5 (12.2%), 2 (4.9%), and 27 (65.9%) patients were diagnosed with complete response (CR), partial response (PR), stable disease (SD), and progressive disease (PD), respectively. The ORR was 29.3%. Immune-related AEs were observed in 35 cases (85.4%), of which 5 cases (12.2%) experienced CTCAE grade $\geq 3$ AEs (Table 3). The most common treatment-related adverse events of any grade were rash (22% of the patients), hypothyroidism (10%), and interstitial pneumonia (10.9%). One patient was diagnosed with severe interstitial pneumonia. He received treatment with hyperbaric oxygen therapy and high-dose steroid administration but died of respiratory failure (grade 5). One patient experienced grade 4 liver dysfunction and received high-dose steroid administration. Two weeks after steroid administration, the levels of hepatic enzyme were normalized.

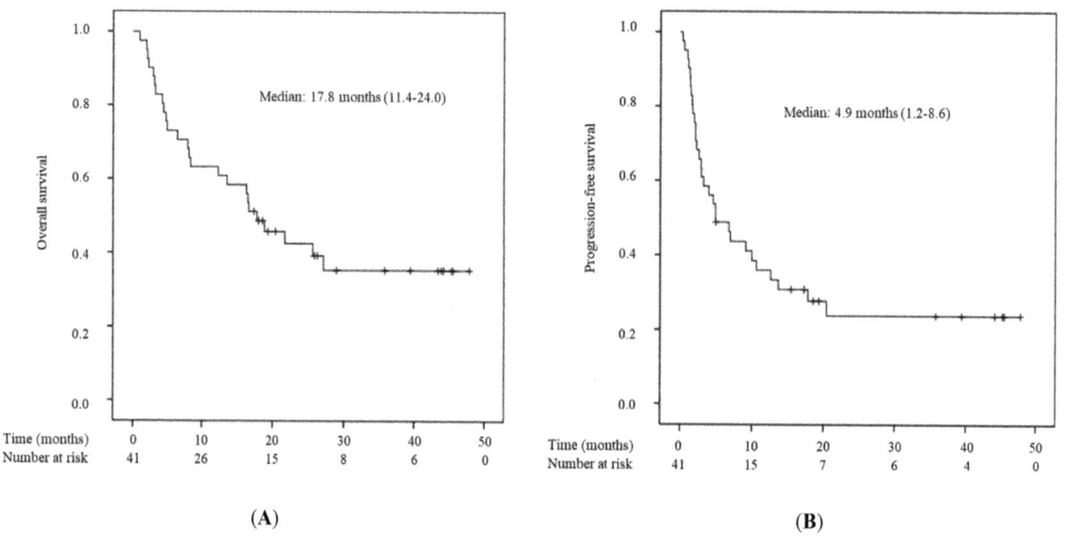


**Figure 1.** A Kaplan–Meier curve of overall survival (**A**) and progression-free survival (**B**) for 41 patients in the cohort study.

**Table 2.** Overall response rate assessed with RECIST version 1.1.

| Overall Response | $n$ = 41 (%) |
|---|---|
| Complete response, $n$ (%) | 7 (17.1) |
| Partial response, $n$ (%) | 5 (12.2) |
| Stable disease, $n$ (%) | 2 (4.88) |
| Progressive disease, $n$ (%) | 27 (65.9) |

**Table 3.** Adverse events in the treated population.

| | Number of Patients (%) | |
|---|---|---|
| | **Any Grade** | **Grade 3, 4, or 5** |
| Any event | 35 (85) | 5 (12) |
| Event leading to treatment discontinuation | 0 | 1 (2) |
| Event leading to death | 0 | 1 (2) |

**Table 3.** *Cont.*

| | Number of Patients (%) | |
|---|---|---|
| | Any Grade | Grade 3, 4, or 5 |
| Infusion reaction | 2 (5) | 2 (5) |
| Interstitial pneumonia | 4(10) | 1 (2) |
| Rash | 9 (22) | |
| Liver dysfunction | 2 (5) | 1 (2) |
| Dysgeusia | 1 (2) | |
| Fatigue | 1 (2) | |
| Hypothyroidism | 4 (10) | |
| Anorexia | 2 (5) | |
| Leg edema | 1 (2) | |
| Adrenal disorder | 2 (5) | 1 (2) |
| Isolated ACTH deficiency | 2 (5) | |
| Parotiditis | 1 (2) | |
| Constipation, diarrhea | 2 (5) | |
| Melena | 1 (2) | |
| Cutaneous sarcoidosis | 1 (2) | |

### *3.3. Risk Factors for Shorter Survival*

As shown in Table 4, univariate analysis showed that ECOG-PS $\geq 2$ ($p = 0.01$), baseline NLR > 3 ($p = 0.01$), hemoglobin $\leq 11$ g/dL ($p = 0.01$), and lower urinary tract tumor ($p = 0.03$) were significantly associated with inferior PFS. However, factors that predict shorter PFS could not be identified based on multivariate analysis. Also, ECOG-PS $\geq 2$ ($p = 0.01$), baseline NLR > 3 ($p = 0.01$), hemoglobin $\leq 11$ g/dL ($p = 0.01$), and CRP $\geq 1$ mg/dL ($p = 0.04$) were significantly associated with inferior OS. Multivariate analysis, including all significant factors identified on univariate analysis, showed that ECOG-PS $\geq 2$ ($p = 0.03$) and NLR > 3 ($p = 0.04$) were significantly associated with inferior OS (Table 5). Antibiotic exposure did not have a significant impact on either PFS or OS.

**Table 4.** Univariate and multivariate Cox's regression analysis of PFS to clinicopathological features in platinum-resistant metastatic urothelial carcinoma patients treated with pembrolizumab.

| | | Univariate Analysis | | Multivariate Analysis | |
|---|---|---|---|---|---|
| **Factor** | **Category** | **HR (95%CI)** | ***p*-Value** | **HR (95%CI)** | ***p*-Value** |
| Age (years) | <72 vs. ≥72 | 1.51 (0.72–3.21) | 0.28 | | |
| Gender | Female vs. Male | 1.07 (0.37–3.09) | 0.90 | | |
| ECOG-PS | 0.1 vs. ≥2 | 5.17 (1.62–16.5) | 0.01 | 2.63 (0.80–8.73) | 0.11 |
| Surgical resection | No vs. Yes | 0.58 (0.28–1.23) | 0.16 | | |
| Any irAEs | Negative vs. Positive | 0.72 (0.35–1.49) | 0.38 | | |
| Neutrophil-to-lymphocyte ratio (NLR) | ≤3.0 vs. >3.0 | 3.01 (1.41–6.42) | 0.01 | 1.97 (0.85–4.57) | 0.12 |
| Hb (g/dL) | ≤11 vs. >11 | 2.70 (1.26–5.75) | 0.01 | 1.89 (0.84–4.57) | 0.13 |
| CRP (mg/dL) | ≤1.0 vs. >1.0 | 1.20 (0.58–2.47) | 0.63 | | |
| Tumor site | Lower vs. Upper | 0.43 (0.20–0.94) | 0.03 | 0.57 (0.25–1.26) | 0.16 |

**Table 4.** *Cont.*

| Factor | Category | Univariate Analysis | | Multivariate Analysis | |
|---|---|---|---|---|---|
| | | HR (95%CI) | *p*-Value | HR (95%CI) | *p*-Value |
| Site of metastasis | Bone | 1.14 (0.49–2.66) | 0.76 | | |
| | Lymph node | 1.07 (0.50–2.29) | 0.87 | | |
| | Lung | 1.07 (0.52–2.18) | 0.86 | | |
| | Liver | 0.89 (0.31–2.57) | 0.83 | | |
| Number of metastases | <1 vs. ≥2 | 1.26 (0.61–2.59) | 0.53 | | |
| Antibiotics prior to pembrolizumab administration | No vs. Yes | 1.16 (0.53–2.54) | 0.71 | | |

**Table 5.** Univariate and multivariate Cox's regression analysis of OS to clinicopathological features in platinum-resistant metastatic urothelial carcinoma patients treated with pembrolizumab.

| Factor | Category | Univariate Analysis | | Multivariate Analysis | |
|---|---|---|---|---|---|
| | | HR (95%CI) | *p*-Value | HR (95%CI) | *p*-Value |
| Age (years) | <72 vs. ≥72 | 2.25 (0.93–5.40) | 0.07 | | |
| Gender | Female vs. Male | 1.19 (0.36–3.98) | 0.78 | | |
| ECOG-PS | 0.1 vs. ≥2 | 20.4 (4.30–96.9) | 0.01 | 6.33 (1.24–32.3) | 0.03 |
| Surgical resection | No vs. Yes | 0.49 (0.22–1.08) | 0.08 | | |
| Any irAEs | Negative vs. Positive | 0.81 (0.37–1.79) | 0.60 | | |
| Neutrophil-to-lymphocyte ratio (NLR) | ≤3.0 vs. >3.0 | 3.53 (1.49–8.36) | 0.01 | 2.79 (1.07–7.23) | 0.04 |
| Hb (g/dL) | ≤11 vs. >11 | 3.38 (1.47–7.95) | 0.01 | 2.35 (0.94–5.90) | 0.07 |
| CRP (mg/dL) | ≤1.0 vs. >1.0 | 2.34 (1.06–5.17) | 0.04 | 2.24 (0.92–5.46) | 0.07 |
| Tumor site | Lower vs. Upper | 0.45 (0.19–1.08) | 0.07 | | |
| Site of metastasis | Bone | 2.37 (0.95–5.76) | 0.06 | | |
| | Lymph node | 1.69 (0.70–4.05) | 0.24 | | |
| | Lung | 1.04 (0.47–2.28) | 0.92 | | |
| | Liver | 1.25 (0.43–3.64) | 0.69 | | |
| Number of metastases | <1 vs. ≥2 | 2.27 (1.01–5.07) | 0.06 | | |
| Antibiotics prior to pembrolizumab administration | No vs. Yes | 1.68 (0.74–3.80) | 0.21 | | |

## 4. Discussion

In this study, pembrolizumab resulted in a median PFS of 4.9 (95% CI, 1.2–8.6) months and a median OS of 17.8 (95% CI, 11.5–24.0) months. These results are slightly better than those of the Keynote-045 trial [4], which reported a median PFS of 2.1 months and a median OS of 10.1 months. In addition, pembrolizumab provided an ORR of 29.3%, higher than in the Keynote-045 trial (21.1%). These results are comparable to those of other retrospective studies evaluating the efficacy and safety of pembrolizumab in advanced UC [26–28].

Several studies proposed biomarkers with prognostic value in patients who received pembrolizumab. In this study, ECOG-PS ≥ 2 and NLR > 3 were identified as independent risk factors for a poor prognosis. In a recent meta-analysis, a poor ECOG-PS, the presence of visceral metastasis, and high pretreatment levels of NLR and CRP were associated with shorter OS [13], similar to the present results. ECOG-PS has been widely used as a tool to validate indications for systemic therapy [14], and poor ECOG-PS has been reported to be associated with shorter OS in patients with advanced UC treated with chemotherapy or ICIs [15,16,29]. Conversely, the phase 2 Keynote-052 trial, which evaluated safety and antitumor activity in patients with locally advanced or metastatic cisplatin-ineligible UC, reported that poor ECOG-PS did not have an impact on the efficacy of pembrolizumab [30]. Parikh et al. reported trends in initiating end-of-life systemic therapy in 1637 patients with metastatic UC [31]. They found a significant increase in the use of systemic therapy in patients with poor PS after the approval of ICIs. The toxicity profile of ICIs compares favorably with chemotherapy, which may have increased the opportunity to administer such drugs to patients with poor PS who would otherwise have been on best supportive care (BSC). Our study also resulted in a higher percentage of patients with ECOG-PS $\geq 2$ (10%) compared with the Keynote-045 trial [4], which included only 0.7% of patients with ECOG-PS $\geq 2$. The reason for this may be that immediately after ICI approval, some patients with poor ECOG-PS who should have been considered for BSC preferred ICI treatment, which has a relatively favorable toxicity profile compared with chemotherapy. However, given the possibility that poor ECOG-PS may lead to shorter OS, whether ICIs should be used in all patients will require further investigation.

The present results also showed that the inflammation-based prognostic marker NLR was significantly associated with OS. There has been controversy over the impact of NLR on the responses to ICI. Previous studies have shown that NLR prior to ICI initiation is associated with survival in patients with metastatic UC [10–13], and that a high NLR is a potential risk factor for poor clinical outcomes in various malignancies [32]. The mechanism by which NLR is related to the response to ICI treatment and survival is uncertain. Kargl et al. reported that a high pretreatment neutrophil count was associated with a decreased number of CD8-positive T cells, which resulted in reduced antitumor activity [33]. Furthermore, Friedlander et al. reported that a higher NLR with shorter survival is due to an increase in the N2 phenotype of the neutrophils, which play a pro-tumorigenic role [34]. Given the potential of NLR as a predictive biomarker that can be easily used in routine clinical practice, further research is warranted to validate its utility and precise mechanism.

Several studies have examined the impact of antibiotic exposure on the response to ICI therapy, but they were inconclusive, with some reporting a positive relationship and others a negative one [17–23,35–37]. In the present study, antibiotic exposure was not related to PFS or OS in patients on ICI treatment. It has been hypothesized that antibiotic exposure may affect responses to ICI treatment through changes in the gut microbiome [37]. Several potential mechanisms explain the relationship between changes in the gut microbiome and ICI response. Mechanisms that include intestinal dendritic cells, which affect T-cell priming and activation, can be affected by specific bacterial strains, and microbiome metabolites modulate host cytokine production and T-cell responses [38,39]. Gopalakrishnan et al. reported interesting findings using a mouse model of melanoma with different compositions of gut microbiota. They found that tumors grew more rapidly in the group with failing microbiota than in those with favorable microbiota, but the tumor growth could be altered by transferring fecal material from the favorable microbiota groups [19]. Despite these findings, antibiotic exposure affected neither PFS nor OS in the present analysis. The relatively small number of patients analyzed in the present study, the definition of the timing, and the duration of antibiotic exposure may be the reasons why no correlation between antibiotic exposure and prognosis was found. In this study, we defined antibiotic exposure as antibiotic use for at least 1 week within 1 month before or after starting ICI treatment with reference to previous reports [40–43]. Reports on the association between the duration of antibiotic exposure and changes in gut microbiome are mixed. Dethlefsen et al. reported the changes in the intestinal microbiota of healthy subjects before and after ciprofloxacin administration. They found that ciprofloxacin had a long-term effect on the post-treatment gut microbiota, but the majority of the gut microbiota returned to pre-treatment levels after 4 weeks [44]. Although antibiotic exposure was not an independent risk factor for poor prognosis in our study, it might be a prognostic factor if the duration of antibiotic exposure were defined differently. Furthermore, another factor responsible for the present result may be the type of antibiotic. Eng et al. reported the impact of antibiotic exposure before ICI treatment on OS in 2737 patients with various types of cancer [45]. They found that antibiotic exposure before ICI treatment was associated with worse OS, but this was observed only with fluoroquinolone exposure and not with penicillin or cephalosporin exposure. Fluoroquinolones can alter many gut microbiota species, including Alistipes, Bifidobacteria, Faecalibacterium, and Ruminococcus, which have been found to affect ICI outcomes [46–50]. In the present study, about 70% of cases used penicillins or cephalosporins, which may be the reason why antibiotic exposure was not identified as a predictor for ICI response. Although several studies have examined the association between ICI treatment outcomes and antibiotic exposure, few have examined this issue specifically in UC patients, and to the best of our knowledge, we are the first to report a lack of correlation between them. To clarify antibiotic exposure as a promising biomarker in real-world clinical practice, it will be necessary to determine whether the ICI response varies depending on the type of antibiotic and the duration of antibiotic exposure in future studies.

In addition to clinicopathological data, multiple biomarker studies have evaluated tumor- and tumor microenvironment-related factors associated with the response to ICI therapy [51]. PD-L1 immunohistochemistry remains a controversial biomarker for ICI treatment. Several single-arm, early phase clinical trials reported PD-L1 expression as a prognostic factor, but not in randomized trials. Several preliminary data in recent clinical trials suggest that patients with a high PD-L1 status have higher ORRs compared with those with a low PD-L1 status [52–54]. On the other hand, the IMvigor211 trial, which assessed the safety and efficacy of atezolizumab (anti-programmed-death-ligand 1 immune checkpoint inhibitor) versus chemotherapy for the treatment of locally advanced or metastatic UC after prior platinum-containing chemotherapy, showed that patients who received atezolizumab lived longer compared with patients who received chemotherapy, regardless of PD-L1 status [55]. In this study, however, PD-L1 immunohistochemistry appeared prognostic but not predictive. Also, the Keynote-045 trial showed that the PD-L1 combined score thresholds of 10% and 1% were not helpful as predictive biomarkers [4]. A systematic review, including 44 trials involving 6664 patients with solid tumors, showed a favorable predictive response of 2.26-fold higher in patients with PD-L1 expression compared with PD-L1-negative patients [56]. To understand the status and perspectives of the predictive response for ICI treatment in UC, three workshops were held from December 2018 to December 2019 [57]. The primary goal of these workshops was to develop recommendations for best approaches to PD-L1 testing in UC. One challenge with the use of PD-L1 immunohistochemistry is the different antibodies and scoring systems used for different agents. This makes the understanding of the role of PD-L1 more complicated, and it will be necessary to establish a uniform measurement system.

Recently, the TMB has been investigated as a promising biomarker to evaluate the response of ICI therapy. Somatic or germline mutations at the DNA level can lead to an increase in tumor-associated antigens, which results in high tumor immunogenicity. Although the relationship between the TMB and the response of ICI therapy has been reported in various types of malignancy, the data regarding UC are not fully elucidated [58–60]. Several studies have examined the impact of the TMB on the response to ICI therapy, but they were inconclusive, with some reporting a positive relationship and others a negative one [52,54]. As a representative study, the IMvigor211 trial showed that patients with a high TMB had longer overall survival in the atezolizumab cohort than the chemotherapy cohort [60]. Conversely, in a cohort of patients treated with atezolizumab in the IMvigor210 trial, there was no relationship between the TMB and clinical benefit [61]. Different methods (assays and cutoffs) may make data interpretation more difficult. Further validation and standardization are needed to elucidate the role of TMB for patients who receive ICI therapy.

Additionally, several studies have focused on the availability of circulating tumor DNA as a biomarker of multiple solid tumors. An analysis of circulating tumor DNA was performed in 29 patients treated with 6 weeks of durvalumab, and a significant reduction in circulating tumor DNA was observed in treatment responders but not in non-responders [7]. Similarly, Vandekerkhove et al. reported that a more aggressive form of disease in 104 patients with metastatic UC showed higher circulating tumor DNA levels [62]. Powles et al. evaluated outcomes in 581 UC patients who were evaluated for circulating tumor DNA from the IMvigor010 trial [63]. They showed that circulating tumor DNA testing at the start of therapy (cycle 1, day 1) identified 37% of the patients who were positive for circulating tumor DNA and who had a poor prognosis. Interestingly, the patients who were positive for circulating tumor DNA had improved disease-free survival and overall survival in the atezolizumab arm versus the observation arm, while no difference was observed in the disease-free survival and overall survival between the treatment arms for circulating tumor DNA negative patients. Further investigations on the role of circulating tumor DNA level as a predictive biomarker in UC patients are required.

The present study has several limitations, including its retrospective nature, the involvement of only two centers, and the small number of patients analyzed. In addition, a significant limitation of the analysis was not performing a shotgun metagenomic analysis of fecal samples to evaluate changes in the intestinal bacterial species and bacterial gene function profiles. Future studies of antibiotic classes, the duration of antibiotic exposure, and analysis of fecal samples are warranted to better understand the association between antibiotic exposure and ICI outcomes. Prospectively validated predictive biomarkers will provide valuable adjuncts to real-world clinical practice, but large trials with longer follow-up will be needed to clarify the many questions remaining.

## 5. Conclusions

In this retrospective study of 41 advanced UC patients receiving ICI treatment, poor ECOG-PS and high NLR were significantly associated with poor prognoses. Antibiotic exposure was not identified as a biomarker for ICI response. To clarify antibiotic exposure as a promising biomarker in real-world clinical practice, it will be necessary to determine whether the ICI response varies depending on the type of antibiotic and the duration of antibiotic exposure in future studies.

**Author Contributions:** Y.A. and Y.Y. proposed and designed the experiments; Y.A., Y.Y. and S.A. performed the experiments; Y.A., M.H., K.K. and A.H. analyzed the data; Y.A. wrote the manuscript; K.F., T.M., K.Y. and H.U. supervised the study. All authors have read and agreed to the published version of the manuscript.

**Funding:** This research received no external funding.

**Institutional Review Board Statement:** This study was approved by the Institutional Review Board of Kindai University Nara Hospital and Kindai University Hospital (Approval number 705).

**Informed Consent Statement:** Informed consent for involvement in it from all of the included patients was waived because of its retrospective design.

**Data Availability Statement:** Data available on request due to privacy or ethical restrictions. The data presented in this study are available on request from the corresponding author. The data are not publicly available due to institutional policy.

**Acknowledgments:** We thank Yuiko Sekine, Shoko Shimizu, and Maiko Maekawa for their assistance in collecting, storing, and managing samples.

**Conflicts of Interest:** H.U. have received honoraria from MSD K.K.

## Abbreviations

| | |
|---|---|
| AEs | Adverse events |
| BSC | best supportive care |
| CD4 | cluster of differentiation 4 |
| CD8 | cluster of differentiation 8 |
| CR | complete response |
| CRP | C-reactive protein |
| CTCAE | common terminology criteria for adverse events |
| DNA | deoxyribonucleic acid |
| ECOG-PS | Eastern Cooperative Oncology Group performance status |
| FN | febrile neutropenia |
| GC | gemcitabine, cisplatin |
| ICI | immune checkpoint inhibitor |

| | |
|---|---|
| IgG4κ | immunoglobulin G4κ |
| MVAC | methotrexate, vinblastine, doxorubicin, and cisplatin |
| NLR | neutrophil-to-lymphocyte ratio |
| ORR | objective response rate |
| OS | overall survival |
| PD | progressive disease |
| PD-L1 | programmed death ligand-1 |
| PFS | progression-free survival |
| PR | partial response |
| PS | performance status |
| RECIST | response evaluation criteria in solid tumors |
| SD | stable disease |
| SPSSs | Statistical Package for the Social Sciences |
| TMB | tumor mutational burden |
| UC | urothelial carcinoma |

## References


1. von der Maase, H.; Sengelov, L.; Roberts, J.T.; Ricci, S.; Dogliotti, L.; Oliver, T.; Moore, M.J.; Zimmermann, A.; Arning, M. Long-Term Survival Results of a Randomized Trial Comparing Gemcitabine Plus Cisplatin, With Methotrexate, Vinblastine, Doxorubicin, Plus Cisplatin in Patients with Bladder Cancer. *J. Clin. Oncol.* **2005**, *23*, 4602–4608. [CrossRef] [PubMed]
2. Keir, M.E.; Butte, M.J.; Freeman, G.J.; Sharpe, A.H. PD-1 and Its Ligands in Tolerance and Immunity. *Annu. Rev. Immunol.* **2008**, *26*, 677–704. [CrossRef]
3. Francisco, L.M.; Salinas, V.H.; Brown, K.E.; Vanguri, V.K.; Freeman, G.J.; Kuchroo, V.K.; Sharpe, A.H. PD-L1 Regulates the Development, Maintenance, and Function of Induced Regulatory T Cells. *J. Exp. Med.* **2009**, *206*, 3015–3029. [CrossRef] [PubMed]
4. Bellmunt, J.; de Wit, R.; Vaughn, D.J.; Fradet, Y.; Lee, J.-L.; Fong, L.; Vogelzang, N.J.; Climent, M.A.; Petrylak, D.P.; Choueiri, T.K.; et al. Pembrolizumab as Second-Line Therapy for Advanced Urothelial Carcinoma. *N. Engl. J. Med.* **2017**, *376*, 1015–1026. [CrossRef] [PubMed]
5. Guidelines Detail. Available online: https://www.nccn.org/guidelines/guidelines-detail (accessed on 4 September 2023).
6. Topalian, S.L.; Taube, J.M.; Anders, R.A.; Pardoll, D.M. Mechanism-Driven Biomarkers to Guide Immune Checkpoint Blockade in Cancer Therapy. *Nat. Rev. Cancer* **2016**, *16*, 275–287. [CrossRef]
7. Kuziora, M.; Higgs, B.W.; Brohawn, P.Z.; Raja, R.; Bais, C.; Ranade, K. Association of Early Reduction in Circulating Tumor DNA (ctDNA) with Improved Progression-Free Survival (PFS) and Overall Survival (OS) of Patients (Pts) with Urothelial Bladder Cancer (UBC) Treated with Durvalumab (D). *J. Clin. Oncol.* **2017**, *35*, 11538. [CrossRef]
8. Yarchoan, M.; Hopkins, A.; Jaffee, E.M. Tumor Mutational Burden and Response Rate to PD-1 Inhibition. *N. Engl. J. Med.* **2017**, *377*, 2500–2501. [CrossRef]
9. Ogihara, K.; Kikuchi, E.; Shigeta, K.; Okabe, T.; Hattori, S.; Yamashita, R.; Yoshimine, S.; Shirotake, S.; Nakazawa, R.; Matsumoto, K.; et al. The Pretreatment Neutrophil-to-Lymphocyte Ratio Is a Novel Biomarker for Predicting Clinical Responses to Pembrolizumab in Platinum-Resistant Metastatic Urothelial Carcinoma Patients. *Urol. Oncol. Semin. Orig. Investig.* **2020**, *38*, 602.e1–602.e10. [CrossRef]
10. Tamura, D.; Jinnouchi, N.; Abe, M.; Ikarashi, D.; Matsuura, T.; Kato, R.; Maekawa, S.; Kato, Y.; Kanehira, M.; Takata, R.; et al. Prognostic Outcomes and Safety in Patients Treated with Pembrolizumab for Advanced Urothelial Carcinoma: Experience in Real-World Clinical Practice. *Int. J. Clin. Oncol.* **2020**, *25*, 899–905. [CrossRef]
11. Yamamoto, Y.; Yatsuda, J.; Shimokawa, M.; Fuji, N.; Aoki, A.; Sakano, S.; Yamamoto, M.; Suga, A.; Tei, Y.; Yoshihiro, S.; et al. Prognostic Value of Pre-Treatment Risk Stratification and Post-Treatment Neutrophil/Lymphocyte Ratio Change for Pembrolizumab in Patients with Advanced Urothelial Carcinoma. *Int. J. Clin. Oncol.* **2021**, *26*, 169–177. [CrossRef]
12. Kobayashi, T.; Ito, K.; Kojima, T.; Kato, M.; Kanda, S.; Hatakeyama, S.; Matsui, Y.; Matsushita, Y.; Naito, S.; Shiga, M.; et al. Risk Stratification for the Prognosis of Patients with Chemoresistant Urothelial Cancer Treated with Pembrolizumab. *Cancer Sci.* **2021**, *112*, 760–773. [CrossRef] [PubMed]
13. Yanagisawa, T.; Mori, K.; Katayama, S.; Mostafaei, H.; Quhal, F.; Laukhtina, E.; Rajwa, P.; Motlagh, R.S.; Aydh, A.; König, F.; et al. Pretreatment Clinical and Hematologic Prognostic Factors of Metastatic Urothelial Carcinoma Treated with Pembrolizumab: A Systematic Review and Meta-Analysis. *Int. J. Clin. Oncol.* **2022**, *27*, 59–71. [CrossRef]
14. Oken, M.M.; Creech, R.H.; Tormey, D.C.; Horton, J.; Davis, T.E.; McFadden, E.T.; Carbone, P.P. Toxicity and Response Criteria of the Eastern Cooperative Oncology Group. *Am. J. Clin. Oncol.* **1982**, *5*, 649–655. [CrossRef]
15. Bajorin, D.F.; Dodd, P.M.; Mazumdar, M.; Fazzari, M.; McCaffrey, J.A.; Scher, H.I.; Herr, H.; Higgins, G.; Boyle, M.G. Long-Term Survival in Metastatic Transitional-Cell Carcinoma and Prognostic Factors Predicting Outcome of Therapy. *J. Clin. Oncol.* **1999**, *17*, 3173–3181. [CrossRef] [PubMed]
16. Khaki, A.R.; Li, A.; Diamantopoulos, L.N.; Bilen, M.A.; Santos, V.; Esther, J.; Morales-Barrera, R.; Devitt, M.; Nelson, A.; Hoimes, C.J.; et al. Impact of Performance Status on Treatment Outcomes: A Real-World Study of Advanced Urothelial Cancer Treated with Immune Checkpoint Inhibitors. *Cancer* **2020**, *126*, 1208–1216. [CrossRef]

17. Ocáriz-Díez, M.; Cruellas, M.; Gascón, M.; Lastra, R.; Martínez-Lostao, L.; Ramírez-Labrada, A.; Paño, J.R.; Sesma, A.; Torres, I.; Yubero, A.; et al. Microbiota and Lung Cancer. Opportunities and Challenges for Improving Immunotherapy Efficacy. *Front. Oncol.* **2020**, *10*, 568939. [CrossRef]
18. Pinato, D.J.; Howlett, S.; Ottaviani, D.; Urus, H.; Patel, A.; Mineo, T.; Brock, C.; Power, D.; Hatcher, O.; Falconer, A.; et al. Association of Prior Antibiotic Treatment with Survival and Response to Immune Checkpoint Inhibitor Therapy in Patients with Cancer. *JAMA Oncol.* **2019**, *5*, 1774–1778. [CrossRef] [PubMed]
19. Gopalakrishnan, V.; Spencer, C.N.; Nezi, L.; Reuben, A.; Andrews, M.C.; Karpinets, T.V.; Prieto, P.A.; Vicente, D.; Hoffman, K.; Wei, S.C.; et al. Gut Microbiome Modulates Response to Anti-PD-1 Immunotherapy in Melanoma Patients. *Science* **2018**, *359*, 97–103. [CrossRef]
20. Matson, V.; Fessler, J.; Bao, R.; Chongsuwat, T.; Zha, Y.; Alegre, M.-L.; Luke, J.J.; Gajewski, T.F. The Commensal Microbiome Is Associated with Anti-PD-1 Efficacy in Metastatic Melanoma Patients. *Science* **2018**, *359*, 104–108. [CrossRef]
21. Derosa, L.; Hellmann, M.D.; Spaziano, M.; Halpenny, D.; Fidelle, M.; Rizvi, H.; Long, N.; Plodkowski, A.J.; Arbour, K.C.; Chaft, J.E.; et al. Negative Association of Antibiotics on Clinical Activity of Immune Checkpoint Inhibitors in Patients with Advanced Renal Cell and Non-Small-Cell Lung Cancer. *Ann. Oncol.* **2018**, *29*, 1437–1444. [CrossRef]
22. Yu, Y.; Zheng, P.; Gao, L.; Li, H.; Tao, P.; Wang, D.; Ding, F.; Shi, Q.; Chen, H. Effects of Antibiotic Use on Outcomes in Cancer Patients Treated Using Immune Checkpoint Inhibitors: A Systematic Review and Meta-Analysis. *J. Immunother.* **2021**, *44*, 76–85. [CrossRef] [PubMed]
23. Kaderbhai, C.; Richard, C.; Fumet, J.D.; Aarnink, A.; Foucher, P.; Coudert, B.; Favier, L.; Lagrange, A.; Limagne, E.; Boidot, R.; et al. Antibiotic Use Does Not Appear to Influence Response to Nivolumab. *Anticancer Res.* **2017**, *37*, 3195–3200. [PubMed]
24. Eisenhauer, E.A.; Therasse, P.; Bogaerts, J.; Schwartz, L.H.; Sargent, D.; Ford, R.; Dancey, J.; Arbuck, S.; Gwyther, S.; Mooney, M.; et al. New Response Evaluation Criteria in Solid Tumours: Revised RECIST Guideline (Version 1.1). *Eur. J. Cancer* **2009**, *45*, 228–247. [CrossRef] [PubMed]
25. Common Terminology Criteria for Adverse Events (CTCAE). Protocol Development CTEP. Available online: https://ctep.cancer.gov/protocolDevelopment/electronic_applications/ctc.htm#ctc_50 (accessed on 4 September 2023).
26. Yasuoka, S.; Yuasa, T.; Nishimura, N.; Ogawa, M.; Komai, Y.; Numao, N.; Yamamoto, S.; Kondo, Y.; Yonese, J. Initial Experience of Pembrolizumab Therapy in Japanese Patients with Metastatic Urothelial Cancer. *Anticancer Res.* **2019**, *39*, 3887–3892. [CrossRef] [PubMed]
27. Swami, U.; Haaland, B.; Kessel, A.; Nussenzveig, R.; Maughan, B.L.; Esther, J.; Sirohi, D.; Pal, S.K.; Grivas, P.; Agarwal, N. Comparative Effectiveness of Immune Checkpoint Inhibitors in Patients with Platinum Refractory Advanced Urothelial Carcinoma. *J. Urol.* **2021**, *205*, 709–717. [CrossRef] [PubMed]
28. Ruiz-Bañobre, J.; Molina-Díaz, A.; Calvo, O.; Fernández, N.; Medina-Colmenero, A.; Santome, L.; Lazaro, M.; Mateos-González, M.; García-Cid, N.; López-López, R.; et al. Rethinking Prognostic Factors in Locally Advanced or Metastatic Urothelial Carcinoma in the Immune Checkpoint Blockade Era: A Multicenter Retrospective Study. *ESMO Open* **2021**, *6*, 100090. [CrossRef] [PubMed]
29. Bellmunt, J.; Choueiri, T.K.; Fougeray, R.; Schutz, F.A.B.; Salhi, Y.; Winquist, E.; Culine, S.; von der Maase, H.; Vaughn, D.J.; Rosenberg, J.E. Prognostic Factors in Patients with Advanced Transitional Cell Carcinoma of the Urothelial Tract Experiencing Treatment Failure with Platinum-Containing Regimens. *J. Clin. Oncol.* **2010**, *28*, 1850–1855. [CrossRef]
30. Balar, A.V.; Castellano, D.; O'Donnell, P.H.; Grivas, P.; Vuky, J.; Powles, T.; Plimack, E.R.; Hahn, N.M.; de Wit, R.; Pang, L.; et al. First-Line Pembrolizumab in Cisplatin-Ineligible Patients with Locally Advanced and Unresectable or Metastatic Urothelial Cancer (KEYNOTE-052): A Multicentre, Single-Arm, Phase 2 Study. *Lancet Oncol.* **2017**, *18*, 1483–1492. [CrossRef]
31. Parikh, R.B.; Galsky, M.D.; Gyawali, B.; Riaz, F.; Kaufmann, T.L.; Cohen, A.B.; Adamson, B.J.S.; Gross, C.P.; Meropol, N.J.; Mamtani, R. Trends in Checkpoint Inhibitor Therapy for Advanced Urothelial Cell Carcinoma at the End of Life: Insights from Real-World Practice. *Oncologist* **2019**, *24*, e397–e399. [CrossRef]
32. Shaul, M.E.; Fridlender, Z.G. Tumour-Associated Neutrophils in Patients with Cancer. *Nat. Rev. Clin. Oncol.* **2019**, *16*, 601–620. [CrossRef]
33. Kargl, J.; Busch, S.E.; Yang, G.H.Y.; Kim, K.-H.; Hanke, M.L.; Metz, H.E.; Hubbard, J.J.; Lee, S.M.; Madtes, D.K.; McIntosh, M.W.; et al. Neutrophils Dominate the Immune Cell Composition in Non-Small Cell Lung Cancer. *Nat. Commun.* **2017**, *8*, 14381. [CrossRef] [PubMed]
34. Fridlender, Z.G.; Albelda, S.M. Tumor-Associated Neutrophils: Friend or Foe? *Carcinogenesis* **2012**, *33*, 949–955. [CrossRef] [PubMed]
35. Sivan, A.; Corrales, L.; Hubert, N.; Williams, J.B.; Aquino-Michaels, K.; Earley, Z.M.; Benyamin, F.W.; Lei, Y.M.; Jabri, B.; Alegre, M.-L.; et al. Commensal Bifidobacterium Promotes Antitumor Immunity and Facilitates Anti–PD-L1 Efficacy. *Science* **2015**, *350*, 1084–1089. [CrossRef] [PubMed]
36. Elkrief, A.; El Raichani, L.; Richard, C.; Messaoudene, M.; Belkaid, W.; Malo, J.; Belanger, K.; Miller, W.; Jamal, R.; Letarte, N.; et al. Antibiotics Are Associated with Decreased Progression-Free Survival of Advanced Melanoma Patients Treated with Immune Checkpoint Inhibitors. *OncoImmunology* **2019**, *8*, e1568812. [CrossRef] [PubMed]
37. Araujo, D.V.; Watson, G.A.; Oliva, M.; Heirali, A.; Coburn, B.; Spreafico, A.; Siu, L.L. Bugs as Drugs: The Role of Microbiome in Cancer Focusing on Immunotherapeutics. *Cancer Treat. Rev.* **2021**, *92*, 102125. [CrossRef]

38. Vétizou, M.; Pitt, J.M.; Daillère, R.; Lepage, P.; Waldschmitt, N.; Flament, C.; Rusakiewicz, S.; Routy, B.; Roberti, M.P.; Duong, C.P.M.; et al. Anticancer Immunotherapy by CTLA-4 Blockade Relies on the Gut Microbiota. *Science* **2015**, *350*, 1079–1084. [CrossRef] [PubMed]
39. Iida, N.; Dzutsev, A.; Stewart, C.A.; Smith, L.; Bouladoux, N.; Weingarten, R.A.; Molina, D.A.; Salcedo, R.; Back, T.; Cramer, S.; et al. Commensal Bacteria Control Cancer Response to Therapy by Modulating the Tumor Microenvironment. *Science* **2013**, *342*, 967–970. [CrossRef] [PubMed]
40. Ishiyama, Y.; Kondo, T.; Nemoto, Y.; Kobari, Y.; Ishihara, H.; Tachibana, H.; Yoshida, K.; Hashimoto, Y.; Takagi, T.; Iizuka, J.; et al. Antibiotic Use and Survival of Patients Receiving Pembrolizumab for Chemotherapy-Resistant Metastatic Urothelial Carcinoma. *Urol. Oncol. Semin. Orig. Investig.* **2021**, *39*, 834.e21–834.e28. [CrossRef]
41. Cortellini, A.; Ricciuti, B.; Facchinetti, F.; Alessi, J.V.M.; Venkatraman, D.; Dall'Olio, F.G.; Cravero, P.; Vaz, V.R.; Ottaviani, D.; Majem, M.; et al. Antibiotic-Exposed Patients with Non-Small-Cell Lung Cancer Preserve Efficacy Outcomes Following First-Line Chemo-Immunotherapy. *Ann. Oncol.* **2021**, *32*, 1391–1399. [CrossRef]
42. Wang, J.-R.; Li, R.-N.; Huang, C.-Y.; Hong, C.; Li, Q.-M.; Zeng, L.; He, J.-Z.; Hu, C.-Y.; Cui, H.; Liu, L.; et al. Impact of Antibiotics on the Efficacy of Immune Checkpoint Inhibitors in the Treatment of Primary Liver Cancer. *Liver Res.* **2022**, *6*, 175–180. [CrossRef]
43. Hopkins, A.M.; Kichenadasse, G.; Karapetis, C.S.; Rowland, A.; Sorich, M.J. Concomitant Antibiotic Use and Survival in Urothelial Carcinoma Treated with Atezolizumab. *Eur. Urol.* **2020**, *78*, 540–543. [CrossRef]
44. Dethlefsen, L.; Huse, S.; Sogin, M.L.; Relman, D.A. The Pervasive Effects of an Antibiotic on the Human Gut Microbiota, as Revealed by Deep 16S rRNA Sequencing. *PLoS Biol.* **2008**, *6*, e280. [CrossRef]
45. Eng, L.; Sutradhar, R.; Niu, Y.; Liu, N.; Liu, Y.; Kaliwal, Y.; Powis, M.L.; Liu, G.; Peppercorn, J.M.; Bedard, P.L.; et al. Impact of Antibiotic Exposure Before Immune Checkpoint Inhibitor Treatment on Overall Survival in Older Adults with Cancer: A Population-Based Study. *J. Clin. Oncol.* **2023**, *41*, 3122–3134. [CrossRef]
46. Routy, B.; Le Chatelier, E.; Derosa, L.; Duong, C.P.M.; Alou, M.T.; Daillère, R.; Fluckiger, A.; Messaoudene, M.; Rauber, C.; Roberti, M.P.; et al. Gut Microbiome Influences Efficacy of PD-1–Based Immunotherapy against Epithelial Tumors. *Science* **2018**, *359*, 91–97. [CrossRef]
47. Lu, P.-H.; Tsai, T.-C.; Chang, J.W.-C.; Deng, S.-T.; Cheng, C.-Y. Association of Prior Fluoroquinolone Treatment with Survival Outcomes of Immune Checkpoint Inhibitors in Asia. *J. Clin. Pharm. Ther.* **2021**, *46*, 408–414. [CrossRef] [PubMed]
48. Stewardson, A.J.; Gaïa, N.; François, P.; Malhotra-Kumar, S.; Delémont, C.; Martinez de Tejada, B.; Schrenzel, J.; Harbarth, S.; Lazarevic, V. Collateral Damage from Oral Ciprofloxacin versus Nitrofurantoin in Outpatients with Urinary Tract Infections: A Culture-Free Analysis of Gut Microbiota. *Clin. Microbiol. Infect.* **2015**, *21*, 344.e1–344.e11. [CrossRef] [PubMed]
49. Rashid, M.-U.; Zaura, E.; Buijs, M.J.; Keijser, B.J.F.; Crielaard, W.; Nord, C.E.; Weintraub, A. Determining the Long-Term Effect of Antibiotic Administration on the Human Normal Intestinal Microbiota Using Culture and Pyrosequencing Methods. *Clin. Infect. Dis.* **2015**, *60*, S77–S84. [CrossRef] [PubMed]
50. Pérez-Cobas, A.E.; Artacho, A.; Knecht, H.; Ferrús, M.L.; Friedrichs, A.; Ott, S.J.; Moya, A.; Latorre, A.; Gosalbes, M.J. Differential Effects of Antibiotic Therapy on the Structure and Function of Human Gut Microbiota. *PLoS ONE* **2013**, *8*, e80201. [CrossRef] [PubMed]
51. Barone, B.; Calogero, A.; Scafuri, L.; Ferro, M.; Lucarelli, G.; Di Zazzo, E.; Sicignano, E.; Falcone, A.; Romano, L.; De Luca, L.; et al. Immune Checkpoint Inhibitors as a Neoadjuvant/Adjuvant Treatment of Muscle-Invasive Bladder Cancer: A Systematic Review. *Cancers* **2022**, *14*, 2545. [CrossRef] [PubMed]
52. Necchi, A.; Anichini, A.; Raggi, D.; Briganti, A.; Massa, S.; Lucianò, R.; Colecchia, M.; Giannatempo, P.; Mortarini, R.; Bianchi, M.; et al. Pembrolizumab as Neoadjuvant Therapy Before Radical Cystectomy in Patients with Muscle-Invasive Urothelial Bladder Carcinoma (PURE-01): An Open-Label, Single-Arm, Phase II Study. *J. Clin. Oncol.* **2018**, *36*, 3353–3360. [CrossRef] [PubMed]
53. van Dijk, N.; Gil-Jimenez, A.; Silina, K.; Hendricksen, K.; Smit, L.A.; de Feijter, J.M.; van Montfoort, M.L.; van Rooijen, C.; Peters, D.; Broeks, A.; et al. Preoperative Ipilimumab plus Nivolumab in Locoregionally Advanced Urothelial Cancer: The NABUCCO Trial. *Nat. Med.* **2020**, *26*, 1839–1844. [CrossRef]
54. Powles, T.; Kockx, M.; Rodriguez-Vida, A.; Duran, I.; Crabb, S.J.; Van Der Heijden, M.S.; Szabados, B.; Pous, A.F.; Gravis, G.; Herranz, U.A.; et al. Clinical Efficacy and Biomarker Analysis of Neoadjuvant Atezolizumab in Operable Urothelial Carcinoma in the ABACUS Trial. *Nat. Med.* **2019**, *25*, 1706–1714. [CrossRef]
55. Van Der Heijden, M.S.; Loriot, Y.; Durán, I.; Ravaud, A.; Retz, M.; Vogelzang, N.J.; Nelson, B.; Wang, J.; Shen, X.; Powles, T. Atezolizumab Versus Chemotherapy in Patients with Platinum-Treated Locally Advanced or Metastatic Urothelial Carcinoma: A Long-Term Overall Survival and Safety Update from the Phase 3 IMvigor211 Clinical Trial. *Eur. Urol.* **2021**, *80*, 7–11. [CrossRef]
56. Khunger, M.; Hernandez, A.V.; Pasupuleti, V.; Rakshit, S.; Pennell, N.A.; Stevenson, J.; Mukhopadhyay, S.; Schalper, K.; Velcheti, V. Programmed Cell Death 1 (PD-1) Ligand (PD-L1) Expression in Solid Tumors as a Predictive Biomarker of Benefit From PD-1/PD-L1 Axis Inhibitors: A Systematic Review and Meta-Analysis. *JCO Precis. Oncol.* **2017**, *1*, 1–15. [CrossRef]
57. Lopez-Beltran, A.; López-Rios, F.; Montironi, R.; Wildsmith, S.; Eckstein, M. Immune Checkpoint Inhibitors in Urothelial Carcinoma: Recommendations for Practical Approaches to PD-L1 and Other Potential Predictive Biomarker Testing. *Cancers* **2021**, *13*, 1424. [CrossRef] [PubMed]
58. Goodman, A.M.; Kato, S.; Bazhenova, L.; Patel, S.P.; Frampton, G.M.; Miller, V.; Stephens, P.J.; Daniels, G.A.; Kurzrock, R. Tumor Mutational Burden as an Independent Predictor of Response to Immunotherapy in Diverse Cancers. *Mol. Cancer Ther.* **2017**, *16*, 2598–2608. [CrossRef] [PubMed]

59. Rosenberg, J.E.; Hoffman-Censits, J.; Powles, T.; van der Heijden, M.S.; Balar, A.V.; Necchi, A.; Dawson, N.; O'Donnell, P.H.; Balmanoukian, A.; Loriot, Y.; et al. Atezolizumab in Patients with Locally Advanced and Metastatic Urothelial Carcinoma Who Have Progressed Following Treatment with Platinum-Based Chemotherapy: A Single-Arm, Multicentre, Phase 2 Trial. *Lancet* **2016**, *387*, 1909–1920. [CrossRef] [PubMed]
60. Powles, T.; Loriot, Y.; Ravaud, A.; Vogelzang, N.J.; Duran, I.; Retz, M.; De Giorgi, U.; Oudard, S.; Bamias, A.; Koeppen, H.; et al. Atezolizumab (Atezo) vs. Chemotherapy (Chemo) in Platinum-Treated Locally Advanced or Metastatic Urothelial Carcinoma (mUC): Immune Biomarkers, Tumor Mutational Burden (TMB), and Clinical Outcomes from the Phase III IMvigor211 Study. *J. Clin. Oncol.* **2018**, *36*, 409. [CrossRef]
61. Snyder, A.; Nathanson, T.; Funt, S.A.; Ahuja, A.; Buros Novik, J.; Hellmann, M.D.; Chang, E.; Aksoy, B.A.; Al-Ahmadie, H.; Yusko, E.; et al. Contribution of Systemic and Somatic Factors to Clinical Response and Resistance to PD-L1 Blockade in Urothelial Cancer: An Exploratory Multi-Omic Analysis. *PLoS Med.* **2017**, *14*, e1002309. [CrossRef]
62. Vandekerkhove, G.; Lavoie, J.-M.; Annala, M.; Murtha, A.J.; Sundahl, N.; Walz, S.; Sano, T.; Taavitsainen, S.; Ritch, E.; Fazli, L.; et al. Plasma ctDNA Is a Tumor Tissue Surrogate and Enables Clinical-Genomic Stratification of Metastatic Bladder Cancer. *Nat. Commun.* **2021**, *12*, 184. [CrossRef]
63. Powles, T.; Assaf, Z.J.; Davarpanah, N.; Banchereau, R.; Szabados, B.E.; Yuen, K.C.; Grivas, P.; Hussain, M.; Oudard, S.; Gschwend, J.E.; et al. ctDNA Guiding Adjuvant Immunotherapy in Urothelial Carcinoma. *Nature* **2021**, *595*, 432–437. [CrossRef] [PubMed]

*cancers*

MDPI

*Article*

# Early Evolution in Cancer: A Mathematical Support for Pathological and Genomic Evidence in Clear Cell Renal Cell Carcinoma

**Annick Laruelle [1,2,*], Claudia Manini [3,4], José I. López [5] and André Rocha [6]**

1 Department of Economic Analysis, University of the Basque Country (UPV/EHU), 48015 Bilbao, Spain
2 IKERBASQUE, Basque Foundation for Science, 48011 Bilbao, Spain
3 Department of Pathology, San Giovanni Bosco Hospital, ASL Città di Torino, 10154 Turin, Italy; claudia.manini@aslcittaditorino.it
4 Department of Sciences of Public Health and Pediatrics, University of Turin, 10124 Turin, Italy
5 Biomarkers in Cancer, Biocruces-Bizkaia Health Research Institute, 48903 Barakaldo, Spain; joseignacio.lopez@biocrucesbizkaia.org
6 Department of Industrial Engineering, Pontifical Catholic University of Rio de Janeiro, Rio de Janeiro CEP22451-900, Brazil; andre-rocha@puc-rio.br
* Correspondence: annick.laruelle@ehu.es

**Simple Summary:** Clear cell renal cell carcinomas (CCRCCs) evolve as dynamic communities of individuals (cells) that are amenable to being studied under sociological rules. Here, the early period of development in CCRCC, progressing from initial homogeneity to high intratumor heterogeneity (ITH) and secondary clonal and sub-clonal diversification, is considered using the hawk-dove game. Fitness is a measure of biological aggressiveness in tumors. The results demonstrate that the fittest clone of a neoplasm in a heterogeneous context is fitter than the clone in a homogeneous environment in the early phases of tumor evolution. This study notes the advantages of a translational multidisciplinary approach in cancer research.

**Abstract:** Clear cell renal cell carcinoma (CCRCC) is an aggressive form of cancer and a paradigmatic example of intratumor heterogeneity (ITH). The hawk-dove game is a mathematical tool designed to analyze competition in biological systems. Using this game, the study reported here analyzes the early phase of CCRCC development, comparing clonal fitness in homogeneous (linear evolutionary) and highly heterogeneous (branching evolutionary) models. Fitness in the analysis is a measure of tumor aggressiveness. The results show that the fittest clone in a heterogeneous environment is fitter than the clone in a homogeneous context in the early phases of tumor evolution. Early and late periods of tumor evolution in CCRCC are also compared. The study shows the convergence of mathematical, histological, and genomics studies with respect to clonal aggressiveness in different periods of the natural history of CCRCC. Such convergence highlights the importance of multidisciplinary approaches for obtaining a better understanding of the intricacies of cancer.

**Keywords:** clear cell renal cell carcinoma; cancer evolution; game theory; intratumor heterogeneity

## 1. Introduction

Renal cell carcinomas rank among the top ten most common neoplasms in Western countries [1] and are a hot topic in modern medicine due to their morphological and genomic variability, complex etiopathogenesis, and resistance to treatment. More specifically, clear cell renal cell carcinoma (CCRCC), an aggressive form of renal cancer accounting for more than 70% of cases [2], currently poses an oncological challenge with promising therapeutic alternatives [3]. CCRCC is a paradigm of intratumor heterogeneity (ITH) and a test bench for new therapies.

Tumor intricacies provide an opportunity to incorporate new scientific approaches. For example, the application of ecological principles to cancer research has led to the coining of the term eco-oncology [4] and to cancer being considered as a social dysfunction [5]. These approaches have brought significant advances in the knowledge of cancer dynamics. As a result, cancer has been found to show at least four different evolutionary pathways: linear, branching, punctuated, and neutral [6]. With the exception of the neutral model, evolution over time is governed by Darwinian principles, where driver mutations generate different clones that make such ITHs unique, unrepeatable tumor cell communities in every case. An exhaustive genomic analysis shows that branching and punctuated models predominate in CCRCC [7].

The assumption that a neoplasm is a huge community of cells interacting with one another enables game theory to be applied to cancer analysis [8]. Many examples of this interdisciplinary cooperation can be found in the literature. In particular, we have recently analyzed the clinical consequences of ITH in CCRCC using the hawk-dove game, showing that the math supports the clinical evidence [9]. Thus, mathematics, histopathology, and genomics come together to demonstrate that tumor aggressiveness is linked to low ITH in the late temporal periods of CCRCC [10].

In continuation with our previous study [10], we focus in this work specifically on the early periods of tumor evolution in CCRCC, providing a mathematical support to the histological and genomic evidence. Coupled with [10], the reader will obtain a complete overview of CCRCC dynamics supported by a mathematical perspective.

## 2. Materials and Methods

### *2.1. Clinical Context*

Large-scale sequencing studies have recently begun to reveal the complexities of this neoplasm. Although most CCRCCs become clinically evident in adulthood (peaking in the 6th–7th decades of life), the first steps of this tumor seem to appear much earlier, in childhood or adolescence, when von Hippel-Lindau (*VHL*) gene mutation happens in one allele of a few hundred cells as a result of chromothripsis [11]. After a dormant period of decades, when the *VHL* gene of the other allele mutates in these patients, the tumor initiates its evolutive course. On that course, up to seven deterministic pathways of temporal and spatial evolution have recently been detected, most of them being either branching- or punctuated-type models that lead to attenuated and accelerated clinical evolution, respectively [7]. However, the same study reveals that a small group of cases displaying only *VHL* gene mutations corresponding to a linear-type evolutionary model show indolent behavior [7]. These genomic data prompted us to investigate whether a mathematical approach would support them.

### *2.2. The Hawk-Dove Game*

Here, we focus on the interactions of cells in three different temporal scenarios of CCRCC. More precisely, we consider cell interactions between elements bearing three well-known genetic driver mutations in CCRCC, i.e., mutations in *VHL*, *PBRM1*, and *BAP-1* genes. *VHL* gene mutation is the common initial step in the vast majority of CCRCCs, and *PBRM1* and *BAP-1* gene mutations are paradigmatic genetic disorders of these tumors with non-aggressive and aggressive behaviors, respectively [12]. For the sake of simplicity, other driver mutations (*SETD2*, etc.) and all of the passenger mutations, which are quite frequent in CCRCC, are not taken into account.

The three temporal scenarios depicted in Figure 1 establish early and late eco-evolutionary periods occurring in many CCRCCs, which correspond roughly to linear, branching, and punctuated temporal models. The late period, i.e., how a CCRCC with high ITH (three cell types, branching model) evolves towards a neoplasm with low ITH (two cell types, punctuated model), has also recently been modeled using the hawk-dove game [9]. Here, however, we focus on modeling the early period, i.e., how homogeneous tumors (one cell type, linear model) evolve into CCRCCs with high ITH (three cell types, branching model).

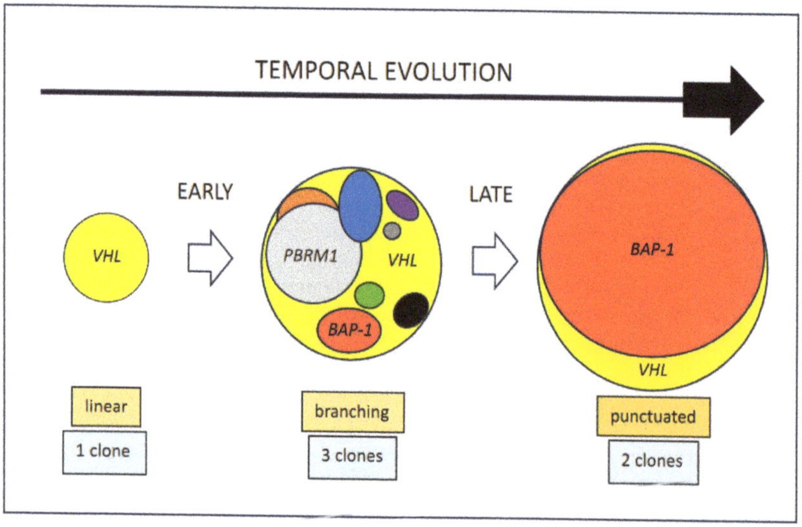


**Figure 1.** Example of a model of temporal evolution in clear cell renal cell carcinomas. The size of the circles reflects tumor aggressiveness (level of clonal fitness). Early evolution exemplifies the transition from linear (homogeneity (*VHL* monodriver clone)) to branching (high heterogeneity (*VHL, PBRM1*, and *BAP-1* driven clones)) models. However, late evolution shows the transition from branching (high heterogeneity (*VHL, PBRM1*, and *BAP-1* driven clones)) to punctuated (low heterogeneity (*VHL, BAP-1* driven clones)) models due to the expansion of an aggressive clone.

The hawk-dove game analyzes competition in biological systems [13]. Here, the game models the bilateral interactions between cells. In each encounter, a cell can behave aggressively, like a hawk, or passively, like a dove, to acquire a resource v. If one cell is aggressive and its opponent is passive, the first obtains the resource and the second gets nothing. If both cells are aggressive, there is a fight and the winner gets the resource, while the loser bears a cost $c > v$. Assuming that they both have the same probability of winning, the expected fitness ("payoff" hereafter) of each cell is $(v - c)/2$. If both cells are passive, one withdraws and gets nothing, while the other takes the resource. Assuming that they both have the same probability of withdrawing, the expected payoff of each cell is $v/2$. These contingencies are summarized in the following payoff matrix:

Let $\alpha$ denote the probability of behaving as a hawk, so that a cell can choose a so-called pure strategy under which it behaves with certainty as either a hawk ($\alpha = 1$) or a dove ($\alpha = 0$). Instead, the cell can choose a mixed strategy ($0 < \alpha < 1$) in which it has a probability $\alpha$ of behaving as a hawk and 1-$\alpha$ of behaving as a dove. In a bilateral encounter, the expected payoff of a cell depends on its own behavior and on that of its opponent. For example, in Table 1, if both cells were to play a mixed strategy $\alpha$, their expected payoffs from such a bilateral encounter would be $\alpha^2(v - c)/2 + \alpha(1 - \alpha)v + 0(1 - \alpha)\alpha + (1 - \alpha)^2v/2$.

**Table 1.** Hawk-dove game payoff matrix.

| | **Hawk** | **Dove** |
|---|---|---|
| Hawk | $(v - c)/2$ | v |
| Dove | 0 | $v/2$ |

In any game, cells that adopt a strategy leading to a higher payoff show both higher fitness and a higher cell replicating ratio. Consequently, the proportion of cells in the population that adopt this advantageous strategy increases over time and may eventually lead to the extinction of cells that adopt other, less fit competing strategies.

The hawk-dove game is studied in three different populations, depending on the temporal scenarios of CCRCC. The population may be homogeneous (linear model), heterogeneous with two types (punctuated model), or heterogeneous with three types (branching model). The types have no intrinsic significance, in the sense that any pair of cells plays the same game (i.e., faces the same payoff matrix as described above). On the other hand, types are important because cells might choose different strategies, leading them to behave differently depending on what type of opponent cell they compete with.

Examples of how tumor cells modify their behavior depending on the environment have been reported previously [14]. When the population is homogeneous (linear model) cells only encounter a single type of cell and can just adopt a single behavior. When the population is heterogeneous, cells may adopt different behaviors when they encounter different types of cells: For instance, they may behave passively when they encounter a certain type of cell and aggressively when they encounter a different type. A cell's payoff in an encounter depends on the behaviors adopted by the cells, and on the proportions of the different types of cells.

We compare the three models on the basis of the payoffs obtained when cells adopt the evolutionarily stable strategy (ESS). The ESS [13] shows the resilience of a given incumbent strategy already being adopted by a population of cells against any other invading strategy in the following sense: Given that mutation ratios are very low during the cell replicating process, consider a population where most members play an ESS while a small proportion of mutants choose a different strategy. In that situation, each mutant's expected payoff is smaller than the expected payoff of a "normal" individual, so mutants are driven out from the population.

An example of the ESS concept applied to the classic hawk-dove game, modeling a homogeneous tumor with only one cell type following an evolutionary linear model, can be found in a benign tumor in which the entire population of cells behave passively and can be invaded by a mutant cell that adopts an aggressive behavior. From Table 1, if a single mutant cell invades the population adopting a hawk strategy, all but one of the incumbent cells will meet another passive cell, leading to an expected payoff $v/2$ for the incumbent cells in each encounter. However, the invading mutant cell will get a payoff $v$ from meeting a passive cell, i.e., a payoff twice as high, giving the cell with the mutation the competitive advantage of replicating faster. In the next period, when cells compete again in such a population, the proportion of passive incumbent cells will have decreased, and that of aggressive mutant cells will have increased. Thus, the dove strategy is not an ESS in such a game, given that it can be invaded. As shown in [13], the ESS in the homogeneous game is the mixed strategy $v/c$: Cells play hawk with a probability (or frequency) of $v/c$ and play dove with a probability of $1 - v/c$. The larger the resource, the more often cells act aggressively, and the larger the cost the more often cells act passively.

In [15], the ESS in heterogeneous populations with two types are studied. The ESS depends on the proportions of the different cells. Some discrimination also arises: One type is always treated better than the other.

Laruelle et al. [9] demonstrate that in heterogeneous populations with three types, there is (under some conditions) an ESS in which each type of cell receives a different treatment.

The comparison of the payoffs in the ESS of the different population games yields the following results (see [9,13,15] and Appendix A for the proofs).

## 3. Results

Result 1: When the population is heterogeneous (i.e., in the punctuated and branching models) one type of cell (the most malignant cells, hereafter "A-cells") obtains a strictly larger payoff than the other in the ESS.

Result 1 suggests that a heterogeneous environment favors discrimination. Cells face the same payoff matrix in all encounters, but the possibility of differentiating cells generates different behaviors when different cells are faced.

Result 2: The payoff that an A-cell obtains in the ESS in a heterogeneous population is strictly larger than that which it would obtain in the ESS in a homogeneous population.

Result 2 suggests that there is an advantage for A-cells in being in a heterogeneous environment rather than a homogeneous one. The intuition is that in homogeneous environments, all cells are treated equally and receive the same expected payoff. In a heterogeneous environment, discrimination arises: Some cells (the A-cells) are treated better than they would be in the homogeneous environment, and others are treated worse. In consequence, A-cells receive a larger payoff than they would in a homogeneous environment.

Result 3: For a given proportion of A-cells, the payoff that an A-cell obtains in the ESS in a punctuated model is strictly larger than it would obtain in the ESS in a branching model (if there is an evolutionarily stable strategy).

Result 3 suggests that there is an advantage for some cells in being in a punctuated environment rather than a branching environment. That is, whenever there is discrimination, it is better for A-cells when there are two groups of cells than when there are three.

## 4. Discussion

Cancer is a disease with a high impact in Western countries, whose complexities defy human understanding. Cell-to-cell interactions govern the temporal and spatial evolution of the disease, which means that every case is unique and unrepeatable. However, certain deterministic evolutionary routes enable future behavioral events to be predicted. Figure 1 illustrates the natural evolution of many CCRCCs.

Mitchell et al. [11] have recently revealed the first steps in the carcinogenesis of these neoplasms, identifying long periods of dormancy prior to the symptomatic phase of the neoplasm. At least theoretically, the initial stages of tumor chronology generate a homogeneous cell growth driven by a *VHL* gene mutation. A few of these tumors will pursue a linear evolution throughout their natural history, the so-called *VHL*-driven CCRCCs, but most will develop subsequent driver mutations, thus generating distinct clones within the tumor [7]. Such new clones (and sub-clones) evolve in a Darwinian model of coexistence, are responsible for ITH, and determine tumor evolution and patient prognosis. VHL syndrome-associated CCRCCs are characteristically multiple and bilateral low-grade neoplasms in which inter-tumor heterogeneity takes place along the natural history. Many CCRCCs are surgically removed from patients during the branching-type period because they usually become symptomatic at that time. Other cases, however, evolve quicker and farther due to the intrinsic aggressiveness of one of these newly generated clones. *BAP-1* gene mutation is a paradigm of these high-fitness clones. Again, following a Darwinian evolutionary pattern, this highly aggressive clone installs the punctuated period in the tumor, dominates subsequent tumor overgrowth, and prompts metastatic development due to local hypoxic pressures in the tumor interior [16].

The later evolutionary step from branching to punctuated model is illustrated in Figure 1. It carries a transition from high to low ITH. Interestingly, the clinical evidence shows that aggressive forms of CCRCC display low ITH at the histological level. Manini et al. [17] have performed a thorough histological analysis of a number of CCRCCs and found that patients with tumors with low variability in the histological grade across the sampling died of the tumor or were symptomatic at last clinical contact, while patients with tumors with a broader spectrum of histological grades were all alive and without clinical disease. Other authors have found that the metastatic ability of a tumor is not always strictly linked to the clone with the highest histological grade [18], so taking grade variations, and not only the highest grade, into consideration may matter in the pathologist's work-up. This simple idea may have promising practical applications for pathologists. Interestingly, genomic analyses also point to the same conclusion. An exhaustive study of more than 1200 tumor regions in 101 CCRCCs has shown that tumors with high chromosomal complexity and low ITH pursue an aggressive clinical course, with early and multiple metastases, while those with low chromosomal complexity and high ITH behave in an attenuated fashion with late, solitary metastases [7]. The conclusion at this point of tumor evolution is that high ITH is linked to an attenuated evolutive course and longer survival than with low ITH tumors. This evidence is supported by a recent mathematical approach using the hawk-dove game [9].

Therapy influences ITH evolution, and this issue deserves comment. Current opinion advocates using the maximum tolerable doses as the preferred strategy, but this forces tumor cells to develop resistances. Antiangiogenic drugs, for example, block neo-angiogenesis in CCRCC, prompting tumor necrosis. However, some tumor cells attempt to generate resistance to such drugs, making it possible for a new tumor made up only of resistant cells to develop. The final result is that maximum tolerable doses turn an initial high ITH (branching-type tumor) into a low ITH (punctuated-type tumor), which is much more aggressive. Promising alternative therapy strategies based on mathematical analyses in the form of the so-called "adaptive therapies" [19] propose the implementation of ecological principles to manage cancer and suggest that the preservation of a high ITH may promote intercellular competition, which could slow down tumor progression. In normal conditions, tumor cells expend their energy on increasing their fitness in an intercellular, competitive way to accomplish all cellular functions. Since energy is limited, a therapeutic strategy using drugs below the maximum tolerable doses will force tumor cells to diversify their energy expenditure across several functions, e.g., increasing fitness and generating resistance, thus slowing down both actions. Interestingly, this double effect means increased survival for patients.

In this paper, however, we focus on the early evolutionary steps of CCRCC from linear to branching model, bringing the transition from homogeneous to high ITH tumors. Our mathematical analysis concludes that high ITH CCRCCs (three cell-type, branching model) are more aggressive than their homogeneous counterparts (one cell-type, linear model). This is the conclusion obtained in the hawk-dove game, and histopathological/genomic studies support it. For example, seven out of 101 CCRCCs in the multi-region genomic analysis performed by Turajlic et al. [7] were made of homogeneous tumor cell populations bearing only *VHL* gene mutations. After the clinical follow-up, this specific subset of tumors behaved indolently and never metastasized [7]. Similarly, the histological analysis of 28 totally sampled CCRCCs by Manini et al. [17] revealed that three of them were totally homogeneous low-grade neoplasms (Grade 1), being alive and without disease after a long-term follow-up. Whether these indolent CCRCCs had evolved over time towards aggressive forms of the disease via subsequent potential clonal/subclonal diversification remains debatable, but possible, and highlights the crucial importance of the exact time of removal in our knowledge of tumor characteristics. A recent review of the importance of ITH in breast, lung, hepatic, and colorectal cancer agrees with this statement [20–27].

The apparent contradiction in the aggressiveness of high ITH is resolved when it is taken into account that cancer is a dynamic disease in which pathologists analyze only static snapshots obtained at the time of tumor removal. That time is variable and not programmable, and depends on multiple clinical factors. A high ITH can thus be better or worse depending on the exact period of tumor evolution and analysis. The natural history of every tumor is not evident in routine practice, and regrettably, current therapies do not factor the evolutionary pattern of tumors into their strategies.

In recent years, game theory has been gaining momentum in oncology. This study notes the importance of translational research in modern medicine and highlights once again the importance of mathematical analyses as an allied tool in cancer analysis [10].

## 5. Conclusions

Here, we consider the initial period in the evolution of many CCRCCs, specifically the time when tumors progress from the original cellular homogeneity towards high ITH as a consequence of clonal and sub-clonal development. Four points derived from our analysis deserve mention: (i) cancer is a dynamic disease in which pathologists analyze a static moment of its evolution; this time is variable and not programmable, (ii) tumors evolve throughout their natural history and may display different degrees of ITH at different times, (iii) this study focuses specifically on the early moments of CCRCC evolution, which is not always paired to clinical findings and diagnosis of the disease, and (iv) the natural history of every tumor is not evident in routine practice and, therefore, not exploitable.

Once more, mathematics supports the clinical evidence and needs to be considered as a tool in the armamentarium for better understanding cancer complexities.

**Author Contributions:** Conceptualization, A.L., C.M., J.I.L. and A.R.; methodology and formal analysis, A.L. and A.R.; clinical investigation, C.M. and J.I.L.; writing—original draft preparation, A.L., C.M., J.I.L. and A.R.; writing—review and editing, A.L., C.M., J.I.L. and A.R. All authors have read and agreed to the published version of the manuscript.

**Funding:** Laruelle acknowledges financial support from grant PID2019-106146GB-I00, funded by MCIN/AEI/10.13039/501100011033, from the ERDF "A Way of Making Europe" scheme and from the Basque Government (Research Group IT1697-22); Rocha acknowledges financial support from the National Council for Scientific and Technological Development—CNPq (CNPq funding 307437/2019-1).

**Institutional Review Board Statement:** Not applicable.

**Informed Consent Statement:** Not applicable.

**Data Availability Statement:** The data presented in this study are available in this article.

**Acknowledgments:** The authors thank Elena Inarra for her fruitful discussion on the model.

**Conflicts of Interest:** The authors declare no conflict of interest.

## Appendix A

We consider three population games. The 1-type population game, denoted by $\Gamma_1(v,c)$, is composed of A-cells; the 2-type population game, denoted by $\Gamma_2(v,c, x_A)$, is composed of A-cells and B-cells; the 3-type population game, denoted by $\Gamma_3(v,c, x_A, x_E)$, is composed of A-cells, B-cells and E-cells. The parameters are v, c, and the proportions $x_I$ of I-cells (I = A, B, E). In each game, a strategy specifies the probability of behaving aggressively when meeting each type of cell present in the population. Let $\alpha_I$ be the probability of behaving aggressively when facing an I-cell (I = A, B, E), and $U_I$ be the payoff of an I-cell. We give the ESS in each of the three games when the A-cells are the cells with the largest payoff, and the E-cells are the cells with the smallest payoff. That is, $U_A > U_B > U_E$.

In game $\Gamma_1(v,c)$, cells only meet A-cells. Strategy $\alpha_A{}^* = v/c$ is the only ESS [13]. The expected payoff of an A-cell in the ESS, denoted by $U_A{}^*$, is given by $U_A{}^* = (1 - v/c)\, v/2$.

In game $\Gamma_2(v,c, x_A)$, a cell meets A-cells and B-cells. The ESS, denoted $(\alpha_A{}^{**}, \alpha_B{}^{**})$, depends on the proportion of A-cells [15]. It is given by

$$(\alpha_A{}^{**}, \alpha_B{}^{**}) = (0, (v/c)(n-1)/(n - nx_A - 1)) \quad \text{if } x_A < z_A$$
$$(0,1) \quad \text{if } z_A < x_A < z_A + 1/n$$
$$(((n-1)v/c - n + nx_A)/(nx_A - 1),1) \quad \text{if } x_A > z_A + 1/n,$$

where $z_A = (1 - v/c)(1 - 1/n)$. The expected payoff of the A-cells in the ESS, denoted by $U_A{}^{**}(x_A)$, is given by:

$$U_A{}^{**}(x_A) = (v/2)(1 - v/c) + (v^2/2c)((2n(1 - x_A) - 1)/(n - nx_A - 1)) \quad \text{if } x_A < z_A$$
$$(v/2)(1 - v/c) + (v/2c)((v(n-1) + cn(1 - x_A))/(n - 1)) \quad \text{if } z_A \leq x_A \leq z_A + 1/n$$
$$(v/2)(1 - v/c) + ((c - v)/c)((vn(1 - x_A))/(nx_A - 1)) \quad \text{if } x_A > z_A + 1/n.$$

The expected payoff of an A-cell peaks when $x_A = z_A$ and starts decreasing thereafter.

In game $\Gamma_3(v,c, x_A, x_E)$, the ESS, denoted $(\alpha_A{}^{***}, \alpha_B{}^{***}, \alpha_E{}^{***})$, exists if and only if $x_A < z_A$ and $x_E < z_E$ [9], where $z_E = (v/c)(1 - 1/n)$. We have $(\alpha_A{}^{***}, \alpha_B{}^{***}, \alpha_E{}^{***}) = (0,((n-1)v/c - nx_E)/(n - nx_A - nx_E - 1),1)$.

The expected payoff of the best-off A-cells in the ESS, denoted by $U_A{}^{***}(x_A,x_E)$, is given by:

$U_A^{***}(x_A,x_E) = v/2(1 - v/c) + v^2/2c((2n(1 - x_A - x_E) - 1)/(n - nx_A - nx_E - 1)) - v/2nx_E/((n - nx_A - nx_E - 1)(n - 1))$.

The larger $x_A$ is, the larger the expected payoff of each A-cell is. By contrast, the larger $x_E$ is, the smaller the expected payoff of each A-cell is.

The first propositions tell us that there is no ESS in games $\Gamma_2(v,c, x_A)$ and $\Gamma_3(v,c, x_A, x_E)$ in which all types of cells obtain the same payoff (Result 1).

**Proposition 1.** *(Iñarra and Laruelle, 2012 [15]) In game $\Gamma_2(v,c, x_A)$, there is no ESS with $U_A = U_B$.*

**Proposition 2.** *In game $\Gamma_3(v,c, xA, xE)$, there is no ESS with UA = UB = UC.*

**Proof.** Suppose that there is an ESS with UA = UB = UC. As shown in [9], the only possible candidate is (v/c, v/c, v/c). If strategy (v/c, v/c, v/c) is not an ESS, there is no possibility to have UA = UB = UC in the ESS. Strategy (v/c, v/c, v/c) is not an ESS if there exists a strategy $(\beta_A, \beta_B, \beta_E)$ such that:

$$u\Big((\frac{v}{c},\frac{v}{c},\frac{v}{c}),(\beta_A,\beta_B,\beta_E)\Big) - u((\beta_A,\beta_B,\beta_E),(\beta_A,\beta_B,\beta_E)) < 0$$

We have

$$u\Big((\frac{v}{c},\frac{v}{c},\frac{v}{c}),(\beta_A,\beta_B,\beta_E)\Big) - u((\beta_A,\beta_B,\beta_E),(\beta_A,\beta_B,\beta_E)) =$$

$$= \sum_{I\in A,B;E} \frac{c}{2}\frac{nx_I}{n-1}\left[\frac{v}{c}\left(1-\frac{1}{n}\right) - \sum_J x_J\beta_J + \frac{\beta_I}{n}\right]\left(\frac{v}{c}-\beta_I\right)$$

Leading to

$$= \frac{c}{2}\frac{n}{n-1}\left[\left(\frac{v}{c}\right)^2\left(\frac{n-1}{n}\right) - 2\frac{v}{c}\left(\frac{n-1}{n}\right)\sum_I x_I\beta_I + \left(\sum_I x_I\beta_I\right)^2 - \frac{1}{n}\left(\sum_I x_I\beta_I^2\right)\right]$$

If one chooses $\beta_A = \frac{v}{c} + \frac{\delta}{x_A}$; $\beta_B = \frac{v}{c} - \frac{\delta}{x_B}$ and $\beta_E = \frac{v}{c}$, we have $0 < \beta_E < 1; \beta_A > 0; \beta_B < 1$ if $\delta > 0$. We also impose $\delta < min\{x_A(1-\frac{v}{c}); \frac{v}{c}x_B\}$ in order to guarantee that $\beta_A < 1$; $\beta_B > 0$.

For those values of $(\beta_A, \beta_B, \beta_E)$, we obtain:

$$u\Big((\frac{v}{c},\frac{v}{c},\frac{v}{c}),(\beta_A,\beta_B,\beta_E)\Big) - u((\beta_A,\beta_B,\beta_E),(\beta_A,\beta_B,\beta_E)) =$$

$$= -\frac{c}{2}\frac{1}{n-1}\left(\frac{\delta^2}{x_A} + \frac{\delta^2}{x_B}\right) < 0.$$

where we used the following:

$$\sum_I x_I\beta_I = \frac{v}{c}\left(\sum_I x_I\right) + \delta - \delta = \frac{v}{c}.$$

Thus, strategy (v/c, v/c, v/c) is not an ESS, and in game $\Gamma_3(v,c,x_A,x_E)$, there is no ESS with $U_A = U_B = U_C$. □

The following propositions indicate when the expected payoff of an A-cell in ESS is the largest in the 2-type population game, then in the 3-type population game, and finally in the 1-type population game: $U_A^{**}(y_A) > U_A^{***}(y_A,x_E) > U_A^*$. That is, $U_A^* < U_A^{***}(y_A,x_E)$ and $U_A^* < U_A^{**}(y_A)$ (Result 2) and $U_A^{**}(y_A) > U_A^{***}(y_A,x_E)$ (Result 3).

**Proposition 3.** *(Laruelle et al., 2023 [9]) Let yA be a proportion of A-cells and let $\Gamma_2$(v,c, yA) and $\Gamma_3$(v,c, yA, xE) with yA < zA and xE < zE. Then, UA**(yA) > UA***(yA, xE).*

**Proposition 4.** *Let $\Gamma_1$(v,c) and $\Gamma_3$(v,c, xA, xE) with xA < zA and xE< zE. Then, UA* < UA*** (xA, xE).*

**Proof.** Function UA***(xA, xE) is increasing with xA and decreasing with xE (see [9], Appendix E). This function reaches its minimum value for xA = 0 and xE = zE. After some algebrics, we obtain UA***(0, zE) = (v/2)(1 − v/c) + (v2/c). This ends the proof, as UA* = (v/2)(1 − (v/c)). □

## References


1. Siegel, R.L.; Miller, K.D.; Wagle, N.S.; Jemal, A. Cancer statistics, 2023. *CA Cancer J. Clin.* **2023**, *73*, 17–48. [CrossRef] [PubMed]
2. Trpkov, K.; Hes, O.; Williamson, S.R.; Adeniran, A.J.; Agaimy, A.; Alaghehbandan, R.; Amin, M.B.; Argani, P.; Chen, Y.B.; Cheng, L.; et al. New developments in existing WHO entities and evolving molecular concepts: The Genitourinary Pathology Society (GUPS) update on renal neoplasia. *Mod. Pathol.* **2021**, *34*, 1392–1424. [CrossRef] [PubMed]
3. Turajlic, S.; Swanton, C.; Boshoff, C. Kidney cancer: The next decade. *J. Exp. Med.* **2018**, *215*, 2477–2479. [CrossRef] [PubMed]
4. Reynolds, B.A.; Oli, M.W.; Oli, M.K. Eco-oncology: Applying ecological principles to understand and manage cancer. *Ecol. Evol.* **2020**, *10*, 8538–8553. [CrossRef] [PubMed]
5. Axelrod, R.; Pienta, K.J. Cancer as a social dysfunction. Why cancer research needs new thinking. *Mol. Cancer Res.* **2018**, *16*, 1346–1347. [CrossRef] [PubMed]
6. Davis, A.; Gao, R.; Navin, N. Tumor evolution: Linear, branching, neutral, or punctuated? *Biochim. Biophys. Acta Rev. Cancer* **2017**, *1867*, 151–161. [CrossRef] [PubMed]
7. Turajlic, S.; Xu, H.; Litchfield, K.; Rowan, A.; Horswell, S.; Chambers, T.; O'Brien, T.; Lopez, J.I.; Watkins, T.B.K.; Nicol, D.; et al. Deterministic evolutionary trajectories influence primary tumor growth: TRACERx Renal. *Cell* **2018**, *173*, 595–610. [CrossRef] [PubMed]
8. Archetti, M.; Pienta, K.J. Cooperation among cancer cells: Applying game theory to cancer. *Nat. Rev. Cancer* **2019**, *19*, 110–117. [CrossRef]
9. Laruelle, A.; Rocha, A.; Manini, C.; López, J.I.; Inarra, E. Effects of heterogeneity on cancer: A game theory perspective. *Bull. Math. Biol.* **2023**, *85*, 72. [CrossRef]
10. Manini, C.; Laruelle, A.; Rocha, A.; López, J.I. Convergent insights into intratumor heterogeneity. *Trends Cancer* **2023**. [CrossRef]
11. Mitchell, T.J.; Turajlic, S.; Rowan, A.; Nicol, D.; Farmery, J.H.R.; O'Brien, T.; Martincorena, I.; Tarpey, P.; Angelopoulos, N.; Yates, L.R.; et al. Timing the landmark events in the evolution of clear cell renal cell cancer: TRACERx Renal. *Cell* **2018**, *173*, 611–623. [CrossRef] [PubMed]
12. Bailey, C.; Black, J.R.M.; Reading, J.L.; Litchfield, K.; Turajlic, S.; McGranahan, N.; Jamal-Hanjani, M.; Swanton, C. Tracking cancer evolution through the disease course. *Cancer Discov.* **2021**, *11*, 916–932. [CrossRef]
13. Maynard-Smith, J.; Price, G.R. The logic of animal conflict. *Nature* **1973**, *246*, 15–18. [CrossRef]
14. Maruyama, T.; Fujita, Y. Cell competition in mammals. Novel homeostatic machinery for embryonic development and cancer prevention. *Curr. Opin. Cell Biol.* **2017**, *48*, 106–112. [CrossRef] [PubMed]
15. Inarra, E.; Laruelle, A. Artificial distinction and real discrimination. *J. Theor. Biol.* **2012**, *305*, 110–117. [CrossRef] [PubMed]
16. Zhao, Y.; Fu, X.; Lopez, J.I.; Rowan, A.; Au, L.; Fendler, A.; Hazell, S.; Xu, H.; Horswell, S.; Shepherd, S.T.C.; et al. Selection of metastasis competent subclones in the tumour interior. *Nat. Ecol. Evol.* **2021**, *5*, 1033–1045. [CrossRef]
17. Manini, C.; López-Fernández, E.; Lawrie, C.H.; Laruelle, A.; Angulo, J.C.; López, J.I. Clear cell renal cell carcinomas with aggressive behavior display low intratumor heterogeneity at the histological level. *Curr. Urol. Rep.* **2022**, *23*, 93–97. [CrossRef]
18. Kim, K.; Zhou, Q.; Christie, A.; Stevens, C.; Ma, Y.; Onabolu, O.; Chintalapati, S.; Mckenzie, T.; Tcheuyap, V.T.; Woolford, L.; et al. Determinants of renal cell carcinoma invasion and metastatic competence. *Nat. Commun.* **2021**, *12*, 5760. [CrossRef]
19. West, J.; Adler, F.; Gallaher, J.; Strobl, M.; Brady-Nicholls, R.; Brown, J.; Roberson-Tessi, M.; Kim, E.; Noble, R.; Viossat, Y.; et al. A survey of open questions in adaptive therapy. Bridging mathematics and clinical translation. *eLife* **2023**, *12*, e84263. [CrossRef]
20. Goyette, M.A.; Lipsyc-Sharf, M.; Polyak, K. Clinical and translational relevance of intratumor heterogeneity. *Trends Cancer* **2023**, *9*, 726. [CrossRef]
21. Zhang, L.L.; Kan, M.; Zhang, M.M.; Yu, S.S.; Xie, H.J.; Gu, Z.H.; Wang, H.N.; Zhao, S.X.; Zhou, G.B.; Song, H.D.; et al. Multiregion sequencing reveals the intratumor heterogeneity of driver mutations in TP53-driven non-small cell lung cancer. *Int. J. Cancer* **2017**, *140*, 103–108. [CrossRef] [PubMed]
22. Zheng, Z.; Yu, T.; Zhao, X.; Gao, X.; Zhao, Y.; Liu, G. Intratumor heterogeneity: A new perspective on colorectal cancer research. *Cancer Med.* **2020**, *9*, 7637–7645. [CrossRef] [PubMed]
23. Guo, M.; Peng, Y.; Gao, A.; Du, C.; Herman, J.G. Epigenetic heterogeneity in cancer. *Biomark. Res.* **2019**, *7*, 23. [CrossRef] [PubMed]

24. Kikutake, C.; Yoshihara, M.; Sato, T.; Saito, D.; Suyama, M. Pan-cancer analysis of intratumor heterogeneity associated with patient prognosis using multidimensional measures. *Oncotarget* **2018**, *9*, 37689–37699. [CrossRef]
25. Dong, L.Q.; Shi, Y.; Ma, L.J.; Yang, L.X.; Wang, X.Y.; Zhang, S.; Wang, Z.C.; Duan, M.; Zhang, Z.; Liu, L.Z.; et al. Spatial and temporal clonal evolution of intrahepatic cholangiocarcinoma. *J. Hepatol.* **2018**, *69*, 89–98. [CrossRef]
26. Swanton, C. Intratumor heterogeneity: Evolution through space and time. *Cancer Res.* **2012**, *72*, 4875–4882. [CrossRef]
27. Morris, L.G.; Riaz, N.; Desrichard, A.; Şenbabaoğlu, Y.; Hakimi, A.A.; Makarov, V.; Reis-Filho, J.S.; Chan, T.A. Pan-cancer analysis of intratumor heterogeneity as a prognostic determinant of survival. *Oncotarget* **2016**, *7*, 10051–10063. [CrossRef]

cancers

Review

# Novel Approaches with HIF-2α Targeted Therapies in Metastatic Renal Cell Carcinoma

**Charles B. Nguyen [1,2,*,†], Eugene Oh [3,†], Piroz Bahar [3], Ulka N. Vaishampayan [1,2], Tobias Else [1,4] and Ajjai S. Alva [1,2]**

1 Rogel Comprehensive Cancer Center, University of Michigan, Ann Arbor, MI 48109, USA; vaishamu@med.umich.edu (U.N.V.); telse@med.umich.edu (T.E.); ajjai@med.umich.edu (A.S.A.)
2 Division of Hematology and Oncology, Department of Internal Medicine, University of Michigan, Ann Arbor, MI 48109, USA
3 University of Michigan Medical School, Ann Arbor, MI 48109, USA; euoh@med.umich.edu (E.O.); baharp@med.umich.edu (P.B.)
4 Division of Metabolism, Endocrinology and Diabetes, Department of Internal Medicine, University of Michigan, Ann Arbor, MI 48109, USA
* Correspondence: ngcharle@med.umich.edu
† These authors contributed equally to this work.

**Simple Summary:** Belzutifan, a hypoxia-inducible factor-2 alpha (HIF-2α) inhibitor, has emerged as an exciting new treatment option not only for patients with Von Hippel-Lindau (VHL)-related renal cell carcinoma (RCC) but also for sporadic RCC. While initial clinical data are promising, potential resistance with HIF-2α inhibitors may occur with increased understanding of this class of therapy. Potential ways to further increase the antitumor activity of HIF-2α targeting include combination strategies with immune checkpoint inhibitors and other targeted agents as well as newer generation HIF-2α inhibitors that are currently under development.

**Abstract:** Germline inactivation of the Von Hippel-Lindau (*VHL*) tumor suppressor is the defining hallmark in hereditary VHL disease and VHL-associated renal cell carcinoma (RCC). However, somatic *VHL* mutations are also observed in patients with sporadic RCC. Loss of function *VHL* mutations result in constitutive activation of hypoxia-inducible factor-2 alpha (HIF-2α), which leads to increased expression of HIF target genes that promote angiogenesis and tumor growth. As of 2023, belzutifan is currently the only approved HIF-2α inhibitor for both VHL-associated and sporadic metastatic RCC (mRCC). However, there is potential for resistance with HIF-2α inhibitors which warrants novel HIF-2α-targeting strategies. In this review, we discuss the potential resistance mechanisms with belzutifan and current clinical trials evaluating novel combinations of belzutifan with other targeted therapies and immune checkpoint inhibitors which may enhance the efficacy of HIF-2α targeting. Lastly, we also discuss newer generation HIF-2α inhibitors that are currently under early investigation and outline future directions and challenges with HIF-2α inhibitors for mRCC.

**Keywords:** renal cell carcinoma; RCC; kidney cancer; belzutifan; HIF-2α targeting; VHL; von Hippel-Lindau

## 1. Introduction

The treatment landscape for metastatic renal cell carcinoma (mRCC) has dramatically changed over the past few decades with the development of targeted therapies as well as immune checkpoint inhibitors (ICI). First-line therapy options for mRCC currently include dual ICI combinations with anti-CTLA-4 and anti-PD-1 inhibitors (e.g., ipilimumab and nivolumab) and tyrosine kinase inhibitors (TKIs), as well as the combination of ICIs with TKIs [1]. Despite these therapies, the estimated 5-year survival rate for patients with mRCC remains low at 15% [2]. Additionally, given the increasing prevalence of ICI-based therapies in the first-line and adjuvant settings in RCC, emerging prospective data have questioned the role of ICI rechallenge in patients with mRCC. In the recent CONTACT-03 trial, patients with mRCC who had disease progression on anti-PD-1-based therapies did not have improved clinical outcomes when immediately re-challenged with atezolizumab, a PD-L1 inhibitor, in combination with cabozantinib [3]. Thus, newer therapeutic targets for mRCC are urgently needed to address these unmet areas.

The *VHL* gene, found on chromosome 3p25, encodes for the VHL tumor suppressor protein that plays pivotal roles in hypoxia response pathway regulation [4]. Under normal oxygen conditions, VHL functions as a ubiquitin ligase that ubiquitinates the alpha subunit of the hypoxia-inducible factor (HIF) transcription factor, resulting in subsequent proteasomal degradation [5]. Inactivating *VHL* mutations lead to HIF stabilization and constitutive activation, resulting in enhanced expression of HIF-regulated genes which all contribute to angiogenesis and tumorigenesis [5–7]. Specifically, HIF induces the expression of *VEGF*, *PDGF-β*, and other angiogenic genes, leading to vascular permeability and endothelial cell differentiation [8]. In addition, HIF stimulates cellular growth by stabilizing and promoting the transcriptional activity of c-Myc which also leads to the enhanced expression of growth signaling genes such as *CCND1* [9]. Cyclin D1, encoded by *CCND1*, ultimately binds to cyclin-dependent kinase 4/6 (CDK4/6), resulting in downstream phosphorylation of the retinoblastoma (RB) tumor suppressor protein and subsequent release and nuclear localization of the E2F transcription factor that drives cell cycle progression [10]. *VHL* inactivation also leads to constitutive phosphorylation and activation of the c-MET receptor tyrosine kinase which interacts with the PI3K/AKT and RAS/MAPK growth pathways, among others [8,11]. HIF contributes to tumor immune escape by directly and indirectly (via c-MET activation) upregulating PD-L1, which inhibits cytotoxic T-cell activation and clonal expansion [12–14]. Additionally, HIF-2α is associated with diminished tumor-infiltrating lymphocytes and enhanced immunosuppressive IL-10 and TGF-β signaling [12]. Germline mutations in *VHL* lead to Von Hippel-Lindau (VHL) disease, an autosomal dominant inherited cancer syndrome. Patients with VHL disease have a near 70% lifetime risk of developing early-onset clear cell RCC (ccRCC), with an average age of RCC diagnosis of approximately 40 years [15]. Somatic mutations in *VHL* are found in about half of sporadic ccRCC cases [5]. Given the emerging data in RCC, the VHL pathway has been explored as a potential therapeutic target in both VHL-associated and sporadic RCC.

As of 2023, belzutifan is the only HIF-2α inhibitor that is Federal Drug Administration (FDA) approved for VHL-associated and sporadic RCC. In the phase 2 MK-6482-004 trial, patients with a confirmed pathogenic germline variant in *VHL* and localized RCC received an oral belzutifan dose of 120 milligrams (mg) daily [16]. At a median follow-up of 21.8 months, the study met the primary endpoint with an objective response rate (ORR) of 49% (95% CI 36–62). The progression-free survival (PFS) rate at 24 months was 96% (95% CI 87–99). Since the study also included patients with VHL-associated RCC and other VHL-associated tumors, responses were also observed in non-RCC lesions, including pancreatic neuroendocrine tumors (ORR 91%) and central nervous system hemangioblastomas (ORR 30%). The most common all-grade adverse events (AEs) with belzutifan in the study included anemia (90%), fatigue (66%), and headaches (41%). Based on the results of this study, the FDA approved belzutifan in August 2021 for the treatment of VHL-associated tumors including RCC [17]. Following this landmark approval for VHL-related tumors, belzutifan has been evaluated in other cancers including sporadic RCC. Ongoing clinical trials of belzutifan in patients with previously treated sporadic mRCC are underway. The LITESPARK-013 trial (NCT04489771) is a phase 2 study comparing two belzutifan doses (120 mg daily vs. 240 mg daily) in patients with sporadic mRCC who previously received up to three lines of prior therapy. The study showed that there was no significant difference in the primary endpoint of ORR as well as PFS and OS between either dose strengths [18]. However, the lower belzutifan dose of 120 mg was associated with less frequent rates of dose modification or discontinuation due to AEs and may represent the optimal starting dose [18]. The LITESPARK-005 trial (NCT04195750) is a randomized, phase 3 trial comparing belzutifan with everolimus, an mTOR inhibitor, in patients with sporadic mRCC who have received up to three lines of prior therapy. In the first interim analysis of 746 patients enrolled, belzutifan led to an improvement in one of the co-primary endpoints of PFS (HR 0.75, 95% CI 0.63–0.90, $p < 0.001$) and the secondary endpoint of ORR (22% vs. 3.5%, $p < 0.00001$), with more frequent complete response (CR) rates compared to everolimus [19]. These preliminary results from LITESPARK-005 led to the recent FDA approval of belzutifan in December 2023 for previously treated sporadic mRCC.

*Resistance Mechanisms of HIF-2α-Targeted Therapies*

Despite the recent advances with belzutifan, the possibility of resistance to HIF-2α inhibition has emerged. While the MK-6482-004 trial, which led to the initial FDA approval of belzutifan in VHL-associated tumors, reported an ORR of 49%, two patients in the study had disease progression with belzutifan, raising the possibility of resistance [16]. Some potential resistance mechanisms to belzutifan and HIF-2α targeting have been proposed. In pre-clinical RCC models, including those with *VHL* alterations, resistance to PT2399 (a preclinical derivative of belzutifan) was associated with lower HIF-2α levels relative to sensitive cells (23% vs. 83%, $p < 0.0001$) but higher expression of *HIF1A* encoding for HIF-1α, which is another HIF that is activated during hypoxic stress [20,21]. RCC tumor xenografts developed resistance to PT2399 when treated for over 100 days, which may be mediated by a G323E mutation in *EPAS1* (endothelial PAS domain-containing protein 1) encoding for HIF-2α [20]. Similarly, in a companion analysis of patients enrolled in a phase 1 study of PT2385, another first-generation HIF-2α inhibitor, patients with disease progression had the *EPAS1* G323E mutation which prevented HIF-2 disassociation, thereby functioning as a "gatekeeper" mutation which may occur upon treatment or at baseline [22]. In the same study, a mutation in *TP53* was also identified at progression, indicating another potential resistance mechanism to HIF-2α-targeted therapies [22]. However, recent work has suggested that intact *TP53* status is not needed for HIF-2α inhibitor sensitivity [23]. Another proposed resistance mechanism with HIF-2α inhibitors includes alterations in the HIF-1/ARNT complex [20,24]. While the precise resistance mechanisms with belzutifan have not been reported, it is possible that similar mechanisms identified in the early generation of HIF-2α-targeted therapies may be involved. Overall, the potential for belzutifan resistance underscores the need for novel HIF-2α-targeting strategies and combinatorial approaches to further enhance efficacy (Table 1).

**Table 1.** Currently active clinical trials of belzutifan in renal cell carcinoma (RCC) as of October 2023.

| Trial Name/NCT Identifier | Phase | Treatment | Setting | Key Primary Endpoint(s) |
|---|---|---|---|---|
| | | Monotherapy trials | | |
| NCT04846920 | 1 | Belzutifan monotherapy (dose escalation) | Advanced refractory ccRCC | Advderse events; percentage of participants who discontinue or modify/interrupt treatment due to adverse event; DLTs |
| LITESPARK-013 (NCT04489771) | 2 | Belzutifan (120 mg versus 240 mg) | Advanced refractory ccRCC | ORR |
| LITESPARK-005 (NCT04195750) | 3 | Belzutifan monotherapy | Advanced refractory ccRCC | PFS and OS (co-primary) |
| | | Combinations with targeted therapies | | |
| LITESPARK-024 (NCT05468697) | 1/2 | Belzutifan + Palbociclib | Advanced refractory ccRCC | Adverse events; DLTs; number of participants who discontinue treatment due to adverse event; ORR (phase 2) |
| NCT04627064 | 1/1B | Belzutifan + Abemaciclib | Advanced refractory ccRCC | Maximum tolerated dose and ORR |
| LITESPARK-003 (NCT03634540) | 2 | Belzutifan + Cabozantinib | Advanced refractory ccRCC | ORR |
| KEYMAKER-U03B (NCT04626518) | 1/2 | Belzutifan + Lenvatinib | Advanced refractory ccRCC | Adverse events; DLTs; number of participants who discontinue therapy due to adverse event; ORR |
| LITESPARK-011 (NCT04586231) | 3 | Belzutifan + Lenvatinib | Advanced refractory ccRCC | PFS and OS (co-primary) |

**Table 1.** *Cont.*

| Trial Name/NCT Identifier | Phase | Treatment | Setting | Key Primary Endpoint(s) |
|---|---|---|---|---|
| | | Combinations with immunotherapy | | |
| LITESPARK-022 (NCT05239728) | 3 | Belzutifan + Pembrolizumab | Adjuvant therapy | Disease-free survival |
| LITESPARK-012 (NCT04736706) | 3 | Belzutifan/Lenvatinib + Pembrolizumab or Quavonlimab | First-line in advanced ccRCC | PFS and OS (co-primary) |
| NCT05899049 (China) | 3 | Belzutifan + Pembrolizumab + Lenvatinib | First-line in advanced ccRCC | PFS and OS (co-primary) |
| NCT05030506 (China) | 1 | Belzutifan + Lenvatinib +/− Pembrolizumab | First-line in advanced ccRCC | Adverse events; DLTs; pharmacokinetic/pharmacodynamic profiles |
| NCT04626479 | 1/2 | Belzutifan + Vibostolimab/Pembrolizumab | First-line in advanced ccRCC | Adverse events; DLTs; ORR |

Abbreviations: ccRCC, clear cell renal cell carcinoma; DLT, dose-limiting toxicities; NCT, National Clinical Trial; ORR, objective response rate; OS, overall survival; PFS, progression-free survival

## 2. Novel Belzutifan Combinations with Targeted Therapies

### 2.1. Belzutifan in Combination with Cabozantinib

Cabozantinib is an oral multi-targeted TKI with activity against VEGFR, c-MET, AXL, and RET that is currently approved for the treatment of metastatic RCC as a monotherapy or in combination with nivolumab [25,26]. HIF-2α notably increases *VEGF* expression and regulates angiogenesis [5,6]. The c-MET (*MET*) receptor tyrosine kinase is activated by the binding of hepatocyte growth factor (HGF), which results in the downstream cell signaling of various pathways involved in cellular growth and migration [27]. As discussed earlier, *VHL* mutations also lead to increased HGF/MET levels in RCC [28]. Thus, concurrent targeting of HIF-2α, VEGFR, and c-MET may be an efficacious approach.

The LITESPARK-003 study (NCT03634540) is an ongoing phase 2 study of belzutifan with cabozantinib in patients with metastatic ccRCC who are treatment-naïve (cohort 1) or have previously received up to two lines of systemic therapy including prior immunotherapy (cohort 2). In this study, patients receive oral belzutifan (120 mg daily) with oral cabozantinib (60 mg daily). In a preliminary analysis of 52 patients enrolled in cohort 2, the primary endpoint of ORR was 30.8% (95% CI 18.7–45.1), including one patient who had a CR and fifteen patients who had partial responses (PRs). Hypertension was the most frequent grade 3–4 AE in 27% of patients [29]. A subsequent update of the LITESPARK-003 study in October 2023 of 50 patients in cohort 1 and the 52 patients in cohort 2 showed an ORR of 70% (95% CI 55–82) and 31% (95% CI 19–45), respectively [30]. The median duration of response was 28.6 months and 31.5 months in cohorts 1 and 2. At a median follow-up of 24.3 and 40 months for cohorts 1 and 2, the reported median PFS was 30.3 months (95% CI 16-not reached [NR]) and 13.8 months (95% CI 9–19). The median OS was not reached in cohort 1 and 26.7 months (95% CI 20–41) in cohort 2 [30]. Taken together, a combination approach with belzutifan and cabozantinib may have synergistic activity by targeting multiple VHL-associated pathways in ccRCC.

### 2.2. Belzutifan in Combination with Lenvatinib

Lenvatinib is a multi-TKI targeting VEGFR1-3, c-Kit, FGFR1-4, PDGR-α, and RET that is currently approved in combination with pembrolizumab for first-line treatment of patients with mRCC and in combination with everolimus for patients who have previously received VEGF-targeted therapy [31–33]. The combination of belzutifan with lenvatinib was previously studied in the B5 arm of the phase 1/2 KEYMAKER-U03B study (NCT04626518). Preliminary data demonstrated an ORR of 50% (95% CI 29–71) in patients with mRCC who were previously treated with immunotherapy and VEGF-TKIs [34]. At a median follow-up after nearly 6 months, the median PFS was 11.2 months (95% CI 4-NR) with a six-month PFS rate of 55%. The most frequent AEs were anemia, fatigue, and hypertension at a frequency of 43% each [34]. The LITESPARK-011 trial (NCT04586231) is an ongoing randomized, phase 3 study of belzutifan with lenvatinib versus cabozantinib monotherapy in patients with mRCC who previously received immunotherapy [35]. The study has co-primary endpoints of PFS per blinded independent central review and overall survival (OS).

### 2.3. Belzutifan in Combination with CDK4/6 Inhibitors

A previously discussed, loss-of-function *VHL* mutations lead to constitutive activation of HIF-2α, resulting in downstream upregulation of *CCND1* encoding for cyclin D1, a cell cycle regulator [7,36]. Cyclin D1 partners with CDK4 or CDK6 to drive cell-cycle progression through phosphorylation and inactivation of the RB tumor suppressor protein [37]. In pre-clinical models, treatment with PT2399 and a CDK4/6 inhibitor resulted in synergistic anti-tumor activity in *VHL*-deficient ccRCC cell cultures and xenografts [38]. Thus, there is potential preclinical rationale for the combination of belzutifan with CDK4/6 inhibitors in metastatic RCC.

Palbociclib is a selective CDK4/6 inhibitor that is currently approved for the treatment of hormone receptor-positive, HER2-negative, advanced breast cancer [39]. The LITESPARK-024 study (NCT05468697) is a randomized phase 1/2 trial comparing the combination of palbociclib with belzutifan versus belzutifan monotherapy in patients with metastatic ccRCC who have been treated with at least two lines of systemic therapy including prior immunotherapy and VEGF-targeted TKIs. The primary endpoint of the phase 2 portion is ORR with key secondary endpoints of OS, PFS, and safety. Abemaciclib is another CDK4/6 inhibitor that is approved for patients with advanced breast cancer and is currently being investigated in combination with belzutifan in a phase 1/1B study (NCT04627064). In this non-randomized study, patients with advanced ccRCC who have received at least one prior VEGF-TKI and one prior ICI will receive abemaciclib alone or in combination with belzutifan. The primary efficacy endpoint for both cohorts is ORR.

## 3. Belzutifan Combinations with Immunotherapy

As discussed earlier, hypoxic conditions not only stimulate HIF signaling but also increase PD-L1 expression with cross-talk between both HIF and tumor immune pathways [13]. Increased HIF-2α levels are associated with diminished numbers of tumor-infiltrating CD8+ lymphocytes and promote stem cell factor (SCF) production, which in turn increases IL-10 and TGF-β secretion that ultimately contributes to an immunosuppressive tumor environment [12]. Thus, there are potential synergistic effects of combining HIF-2α inhibitors with ICIs.

### 3.1. Triplet Combinations with Belzutifan and Immunotherapy

In recent years, active clinical investigation in mRCC has moved from doublet-based strategies (e.g., dual ICI regimens and ICI-TKI combinations) to novel triplet therapy combinations. The COSMIC-313 trial was the first phase 3 study to evaluate a triplet-based combination in mRCC [40]. In this study, patients with untreated, intermediate/poor-risk mRCC were treated with the triplet combination of ipilimumab/nivolumab and cabozantinib versus ipilimumab/nivolumab and placebo. The study met the primary endpoint of PFS (HR 0.73, 95% CI 0.57–0.94, $p = 0.01$) in favor of ipilimumab/nivolumab and cabozantinib; however, the triplet regimen was notably associated with more frequent grade 3–4 AEs (79%) compared to the control group (59%) [40]. Additionally, no OS data have yet been reported, and there is a question as to why patients with poor-risk disease did not derive benefit from triplet therapy in the subgroup analysis compared to patients with intermediate-risk disease. Nonetheless, this trial was a novel study of triplet therapy in mRCC.

An ongoing trial is evaluating the triplet combination of belzutifan with lenvatinib and pembrolizumab (NCT04736706). This phase 3 study will randomize patients with untreated metastatic ccRCC to either belzutifan/lenvatinib/pembrolizumab, belzutifan/lenvatinib with quavonlimab (anti-CTLA-4 ICI), or lenvatinib/pembrolizumab [41]. The co-primary endpoints for this trial are PFS and OS with key secondary endpoints of ORR and toxicity.

Similar trials with the triplet combination of belzutifan with lenvatinib and pembrolizumab are also ongoing (NCT05899049 and NCT05030506).

### 3.2. Belzutifan and Immunotherapy Combinations in the Adjuvant Setting

The data supporting the use of adjuvant treatment for localized RCC have been mixed. Two studies evaluating the role of adjuvant sunitinib (S-TRAC and ASSURE trials) demonstrated conflicting results regarding the primary endpoint of disease-free survival (DFS) in patients with high-risk, localized RCC, although potential differences in patient selection in both studies could have explained the contradictory outcomes [42,43]. In the current immunotherapy era, pembrolizumab is the only ICI approved for adjuvant therapy in patients with ccRCC who have a high risk of recurrence following resection based on the results of the KEYNOTE-564 trial. This study demonstrated an improvement in DFS with pembrolizumab compared to placebo [44]. Interestingly, trials of other ICI agents in the adjuvant setting have yielded negative results [45,46]. Nonetheless, building on the DFS benefit of adjuvant pembrolizumab, the LITESPARK-022 trial (NCT05239728) is an ongoing phase 3 study evaluating the combination of pembrolizumab and belzutifan versus pembrolizumab and placebo as adjuvant therapy in patients with ccRCC following nephrectomy or metastasectomy who are at high risk of disease recurrence [47]. The study's primary endpoint is DFS with secondary endpoints including OS and safety.

### 3.3. Belzutifan Combination with Anti-TIGIT Therapies

An emerging ICI target is a T cell immunoreceptor with immunoglobulin and an ITIM domain (TIGIT) which is involved in immune-mediated tumor recognition [48]. Early phase studies evaluating anti-TIGIT antibodies have demonstrated potential antitumor activity in a subset of patients with metastatic solid tumors [49–51]. The combination of vibostolimab, a novel anti-TIGIT agent, with pembrolizumab was investigated in a phase 1 study which showed an ORR of 26% in patients with advanced non-small cell lung cancer who have not received ICIs [52]. There is an ongoing phase 1b/2 study of a proprietary co-formulation of vibostolimab and pembrolizumab with the addition of belzutifan as a first-line therapy for patients with mRCC (NCT04626479).

## 4. Other Novel HIF-Targeted Therapies in RCC

The clinical success of belzutifan has paved the way for other newer generation HIF-2α-targeted agents in the drug development pipeline. There are both pre-clinical and clinical-level agents differentiated from belzutifan that are currently in development for RCC, with some demonstrating early promise for improved therapeutic outcomes (Table 2).

**Table 2.** Currently active clinical trials of newer generation HIF-2α inhibitors for mRCC as of October 2023.

| Trial Name/NCT Identifier | Phase | Agent | Setting | Key Primary Endpoint(s) |
|---|---|---|---|---|
| | | Novel small molecule inhibitors | | |
| NCT04895748 | 1/1b | DFF332 monotherapy<br>DFF332 + Everolimus/Spartalizumab + Taminadenant | Advanced refractory ccRCC | Adverse events; DLTs; number of participants with dose interruptions/reductions; dose intensity for dose escalation/expansion |
| NCT05119335 | 1/2 | NKT-2152 | Advanced refractory ccRCC | DLTs; recommended dose for expansion in the dose escalation phase (phase 1); recommended dose for phase 2; ORR (phase 2) |
| NCT05935748 | 2 | NKT-2152 + Palbociclib + Sasanlimab | Advanced refractory ccRCC | DLT and ORR |
| ARC-20 (NCT05536141) | 1 | AB521 | Advanced refractory ccRCC | DLTs and adverse events |
| NCT05843305 | 1 | BPI-452080 | Advanced refractory ccRCC | Adverse events |
| | | RNA interference (RNAi) | | |
| NCT04169711 | 1 | ARO-HIF2 | Advanced refractory ccRCC | Adverse events |

Abbreviations: ccRCC, clear cell renal cell carcinoma; DLT, dose-limiting toxicities; NCT, National Clinical Trial; ORR, objective response rate

### 4.1. DFF332

DFF332 is a small-molecule HIF-2α inhibitor that is being studied in a phase 1/1b study as a monotherapy and in combination with everolimus (an mTOR inhibitor) or spartalizumab (an anti-PD-1 ICI), plus taminadenant (an adenosine A2a receptor antagonist) in patients with mRCC and other tumors harboring HIF alterations (NCT04895748). The primary endpoints for this dose escalation and expansion study include safety and frequency of dose-limiting toxicities, with key secondary endpoints of ORR and PFS [53].

### 4.2. ARO-HIF2 (RNA Interference)

Targeted RNA silencing or interference of gene expression holds promise as a novel therapeutic mechanism. In pre-clinical models, novel small interfering RNA (siRNA) against HIF-2α such as ARO-HIF2 were shown to decrease HIF-2α levels and tumor volume in ccRCC tumorgraft models [54]. More recently, ARO-HIF2 is being evaluated in a phase 1 study of patients with previously treated mRCC (NCT04169711). This study has a primary endpoint of safety and secondary endpoints including tumor response. In an interim analysis of the first 23 patients enrolled, fatigue was the most frequent AE reported in 39% of patients [55]. Serious AEs were reported in three patients, including hypoxia, myocarditis, and neuropathy. Efficacy analysis showed a disease control rate (CR, PR, and stable disease) of 30% [55]. Correlative analysis of enrolled patients who underwent on-treatment biopsies showed that ARO-HIF2 led to reductions in HIF-2α mRNA and protein levels. Overall, this initial data showed proof-of-concept of RNA-based therapies targeting HIF-2α expression, although further clinical investigation is needed to elucidate its efficacy.

### 4.3. NKT-2152

NKT-2152 is a novel small-molecule HIF-2α inhibitor under investigation. In cell culture and mouse models, NKT-2152 was shown to interfere with HIF-2α degradation and disrupt the HIF-2α/HIF-1β complex, leading to decreased nuclear localization of HIF-2α and subsequent expression of HIF-2α target genes such as *VEGF-A*, *GLUT1*, and *CCND1* (cyclin D1) [56]. A phase 1/2 trial (NCT05119335) of NKT-2152 is currently ongoing in patients with previously treated mRCC with primary endpoints of identifying the recommended dose for expansion (phase 1 portion) and ORR (phase 2 portion). Another phase 2 study (NCT05935748) is investigating the combination of NKT-2152 with palbociclib, a CDK4/6 inhibitor, with or without sasanlimab, a novel anti-PD-1 ICI, in patients with previously treated mRCC. The primary endpoints of this study include the frequency of dose-limiting toxicities and ORR.

### 4.4. AB521

Another novel HIF-2α inhibitor identified is AB521, which inhibits HIF-2α binding to HIF-1β and therefore disrupts expression of HIF-2α target genes, which has been investigated in pre-clinical models [57]. In RCC tumor xenograft models, treatment with AB521 resulted in a dose-dependent reduction in tumor size [58]. Furthermore, the combination of AB521 with cabozantinib resulted in potential synergistic activity in pre-clinical models compared to either agent alone [59]. In a pharmacokinetic and pharmacodynamic phase 1 study, AB521 was shown to decrease circulating levels of erythropoietin (EPO) in a dose-dependent manner [60]. There is currently an ongoing phase 1 trial (NCT05536141) that is investigating the safety and efficacy of AB521 in patients with mRCC who have previously received anti-PD-1 or TKI therapy, as well as in patients with other advanced solid tumors.

### 4.5. BPI-452080

BPI-452080 is a selective small-molecule HIF-2α inhibitor that disrupts HIF-2α heterodimerization with HIF-1β, leading to decreased transcription of downstream hypoxia-responsive genes (e.g., *GLUT1*, *CCND1*, and *CXCR4*) and VEGFA secretion in pre-clinical RCC models [61]. In RCC tumor xenograft models, oral treatment with BPI-452080 led to a dose-dependent reduction in tumor size [61]. As of early 2023, a phase 1 clinical trial of BPI-452080 (NCT05843305) is ongoing in China to evaluate the safety and efficacy in patients with mRCC (including VHL-associated mRCC) and other advanced solid tumors.

### 4.6. KD061

Ferroptosis is the process of cellular accumulation of cytotoxic iron-dependent lipid peroxides that results in programmed cell death, in a manner that is distinct from apoptosis and autophagy, and may be a potential target in mRCC [62]. KD061 is a recently identified molecule that targets iron-sulfur cluster assembly 2 (ISCA2), which is involved in ferroptosis and also decreases HIF-1/2α levels [63]. In pre-clinical RCC models, treatment with KD061 resulted in decreased tumor growth and HIF levels as well as induced ferroptosis [63]. Thus, dual targeting of HIF and ferroptotic pathways may overcome therapeutic resistance and further increase the efficacy of HIF-targeted inhibition. It will be interesting to see if KD061 will transition to first-in-human early-phase clinical studies in the future.

## 5. Broadening the Use of HIF-2α-Targeted Therapies to Other Cancers

While HIF-2α clearly plays critical roles in ccRCC, there is emerging data that HIF-2α may also be implicated in other cancers such as breast, colorectal, liver, and prostate cancers [64–67]. A few prospective studies evaluating the efficacy of belzutifan in other cancers beyond mRCC are currently ongoing. MK-6482-015 is a phase 2 study (NCT04924075) evaluating the efficacy of belzutifan in patients with advanced pheochromocytoma/paraganglioma, neuroendocrine tumors of the pancreas, and gastrointestinal stromal tumors (GIST), among others. Belzutifan is also being studied in combination with pembrolizumab in patients with metastatic castration-resistant prostate cancer (NCT02861573) as well as in the neoadjuvant setting in combination with abiraterone acetate, prednisone, and leuprolide acetate in patients with regional node-positive prostate cancer (NCT05574712).

## 6. Conclusions and Future Directions

HIF-2α targeting with belzutifan holds exciting promise and is a welcomed therapeutic addition to the evolving treatment landscape for patients with VHL-associated and sporadic mRCC. Since most first-line systemic therapy options for mRCC largely include immunotherapy-based combinations, HIF-2α inhibitors fulfill the need for newer therapies with novel mechanisms of action for patients who have previously been treated with ICIs, although clinical trials of HIF-2α inhibitors in the first-line setting are ongoing. As the treatment of mRCC now enters the era of HIF-2α-targeted therapies, an understanding and recognition of the potential resistance to HIF-2α inhibitors that may emerge with longer follow-up will be critical moving forward. Potential combinatorial strategies with HIF-2α agents and immunotherapy and other targeted therapies may further enhance the efficacy of HIF-2α inhibitors. Other newer generation HIF-2α inhibitors are also under early phase clinical investigation, which may lead to additional therapy options for patients with mRCC. It is currently unknown if newer generation HIF-targeted agents can overcome the *EPAS1* G323E resistance mutation. Future HIF-2α inhibitors should be designed to potentially target this "gatekeeper" G323E mutation and other mediators of resistance within the VHL/HIF pathway. An additional potential challenge that could emerge is determining the optimal sequencing of therapies, including HIF-2α inhibitors. Future prospective studies should evaluate if HIF-2α-targeted therapies and combinations have the greatest clinical benefit either in the first-line or subsequent settings, as well as investigate if HIF-2α inhibitors can be continued or re-challenged at progression. Currently, there are no predictive biomarkers in clinical practice to select therapies such as immune-based or TKI-based regimens for patients with mRCC. While *VHL* alterations are frequent in mRCC, not all patients with mRCC will harbor *VHL*-inactivating mutations [4,68]. Thus, an enhanced selection of patients with mRCC who have VHL pathway alterations or HIF-sensitizing mutations by incorporating biomarker analysis may be needed to identify the subgroup of patients that may or may not respond to HIF-2α inhibitors.

**Author Contributions:** Conceptualization, C.B.N., E.O. and A.S.A.; writing—original draft preparation, C.B.N., E.O. and P.B.; writing—review and editing, C.B.N., E.O., P.B., U.N.V., T.E. and A.S.A.; supervision, U.N.V., T.E. and A.S.A. All authors have read and agreed to the published version of the manuscript.

**Funding:** This research received no external funding.

**Conflicts of Interest:** Ajjai S. Alva has served as an advisor to AstraZeneca, Bristol-Myers Squibb, Merck, and Pfizer. He has previously received institutional research funding from Arcus Biosciences, Astellas, AstraZeneca, Bayer, Bristol-Myers Squibb, Celgene, Genentech, Merck Sharp & Dohme, Janssen, Mirati Therapeutics, Progenics, Prometheus Laboratories, and Roche. He has also received travel and accommodation from Bristol-Myers Squibb and Merck. Tobias Else has served as an advisor to Corcept Therapeutics, HRA Pharmaceuticals, Merck, and Lantheus. Ulka N. Vaishampayan has served as advisor to Alkermes, Aveo, Bayer, Bristol-Myers Squibb/Medarex, Exelixis, Gilead, Merck, EMD Serono, Pfizer, and Sanofi. She has received research support from Bristol-Myers Squibb, Exelixis, and Merck. She has had speaking activities with Bayer, Exelixis, and Astellas. All other co-authors have no disclosures to report.

## References

1. Aldin, A.; Besiroglu, B.; Adams, A.; Monsef, I.; Piechotta, V.; Tomlinson, E.; Hornbach, C.; Dressen, N.; Goldkuhle, M.; Maisch, P.; et al. First-line therapy for adults with advanced renal cell carcinoma: A systematic review and network meta-analysis. *Cochrane Database Syst. Rev.* **2023**, *5*, CD013798. [CrossRef]
2. Siegel, R.L.; Miller, K.D.; Wagle, N.S.; Jemal, A. Cancer statistics, 2023. *CA Cancer J. Clin.* **2023**, *73*, 17–48. [CrossRef]
3. Pal, S.K.; Albiges, L.; Tomczak, P.; Suarez, C.; Voss, M.H.; de Velasco, G.; Chahoud, J.; Mochalova, A.; Procopio, G.; Mahammedi, H.; et al. Atezolizumab plus cabozantinib versus cabozantinib monotherapy for patients with renal cell carcinoma after progression with previous immune checkpoint inhibitor treatment (CONTACT-03): A multicentre, randomised, open-label, phase 3 trial. *Lancet* **2023**, *402*, 185–195. [CrossRef]
4. Cowey, C.L.; Rathmell, W.K. VHL gene mutations in renal cell carcinoma: Role as a biomarker of disease outcome and drug efficacy. *Curr. Oncol. Rep.* **2009**, *11*, 94–101. [CrossRef]
5. Kim, W.Y.; Kaelin, W.G. Role of VHL gene mutation in human cancer. *J. Clin. Oncol.* **2004**, *22*, 4991–5004. [CrossRef]
6. Choueiri, T.K.; Kaelin, W.G., Jr. Targeting the HIF2-VEGF axis in renal cell carcinoma. *Nat. Med.* **2020**, *26*, 1519–1530. [CrossRef]
7. Zatyka, M.; da Silva, N.F.; Clifford, S.C.; Morris, M.R.; Wiesener, M.S.; Eckardt, K.U.; Houlston, R.S.; Richards, F.M.; Latif, F.; Maher, E.R. Identification of cyclin D1 and other novel targets for the von Hippel-Lindau tumor suppressor gene by expression array analysis and investigation of cyclin D1 genotype as a modifier in von Hippel-Lindau disease. *Cancer Res.* **2002**, *62*, 3803–3811.
8. Mazumder, S.; Higgins, P.J.; Samarakoon, R. Downstream Targets of VHL/HIF-alpha Signaling in Renal Clear Cell Carcinoma Progression: Mechanisms and Therapeutic Relevance. *Cancers* **2023**, *15*, 1316. [CrossRef]
9. Gordan, J.D.; Bertout, J.A.; Hu, C.J.; Diehl, J.A.; Simon, M.C. HIF-2alpha promotes hypoxic cell proliferation by enhancing c-myc transcriptional activity. *Cancer Cell* **2007**, *11*, 335–347. [CrossRef]
10. Sager, R.A.; Backe, S.J.; Ahanin, E.; Smith, G.; Nsouli, I.; Woodford, M.R.; Bratslavsky, G.; Bourboulia, D.; Mollapour, M. Therapeutic potential of CDK4/6 inhibitors in renal cell carcinoma. *Nat. Rev. Urol.* **2022**, *19*, 305–320. [CrossRef] [PubMed]
11. Nakaigawa, N.; Yao, M.; Baba, M.; Kato, S.; Kishida, T.; Hattori, K.; Nagashima, Y.; Kubota, Y. Inactivation of von Hippel-Lindau gene induces constitutive phosphorylation of MET protein in clear cell renal carcinoma. *Cancer Res.* **2006**, *66*, 3699–3705. [CrossRef]
12. Wu, Q.; You, L.; Nepovimova, E.; Heger, Z.; Wu, W.; Kuca, K.; Adam, V. Hypoxia-inducible factors: Master regulators of hypoxic tumor immune escape. *J. Hematol. Oncol.* **2022**, *15*, 77. [CrossRef]
13. Shurin, M.R.; Umansky, V. Cross-talk between HIF and PD-1/PD-L1 pathways in carcinogenesis and therapy. *J. Clin. Investig.* **2022**, *132*, e159473. [CrossRef] [PubMed]
14. Kammerer-Jacquet, S.F.; Medane, S.; Bensalah, K.; Bernhard, J.C.; Yacoub, M.; Dupuis, F.; Ravaud, A.; Verhoest, G.; Mathieu, R.; Peyronnet, B.; et al. Correlation of c-MET Expression with PD-L1 Expression in Metastatic Clear Cell Renal Cell Carcinoma Treated by Sunitinib First-Line Therapy. *Target. Oncol.* **2017**, *12*, 487–494. [CrossRef]
15. Maher, E.R.; Neumann, H.P.; Richard, S. von Hippel-Lindau disease: A clinical and scientific review. *Eur. J. Hum. Genet.* **2011**, *19*, 617–623. [CrossRef]
16. Jonasch, E.; Donskov, F.; Iliopoulos, O.; Rathmell, W.K.; Narayan, V.K.; Maughan, B.L.; Oudard, S.; Else, T.; Maranchie, J.K.; Welsh, S.J.; et al. Belzutifan for Renal Cell Carcinoma in von Hippel-Lindau Disease. *N. Engl. J. Med.* **2021**, *385*, 2036–2046. [CrossRef]
17. Fallah, J.; Brave, M.H.; Weinstock, C.; Mehta, G.U.; Bradford, D.; Gittleman, H.; Bloomquist, E.W.; Charlab, R.; Hamed, S.S.; Miller, C.P.; et al. FDA Approval Summary: Belzutifan for von Hippel-Lindau Disease-Associated Tumors. *Clin. Cancer Res.* **2022**, *28*, 4843–4848. [CrossRef]
18. Agarwal, N.; Brugarolas, J.; Ghatalia, P.; George, S.; Haanen, J.B.A.G.; Gurney, H.P.; Ravilla, R.; Van der Veldt, A.A.M.; Beuselinck, B.; Pokataev, I.; et al. 1881O Safety and efficacy of two doses of belzutifan in patients (pts) with advanced RCC: Results of the randomized phase II LITESPARK-013 study. *Ann. Oncol.* **2023**, *34*, S1011. [CrossRef]

19. Albiges, L.; Rini, B.I.; Peltola, K.; De Velasco Oria, G.A.; Burotto, M.; Suarez Rodriguez, C.; Ghatalia, P.; Lacovelli, R.; Lam, E.T.; Verzoni, E.; et al. LBA88 Belzutifan versus everolimus in participants (pts) with previously treated advanced clear cell renal cell carcinoma (ccRCC): Randomized open-label phase III LITESPARK-005 study. *Ann. Oncol.* **2023**, *34*, S1329–S1330. [CrossRef]
20. Chen, W.; Hill, H.; Christie, A.; Kim, M.S.; Holloman, E.; Pavia-Jimenez, A.; Homayoun, F.; Ma, Y.; Patel, N.; Yell, P.; et al. Targeting renal cell carcinoma with a HIF-2 antagonist. *Nature* **2016**, *539*, 112–117. [CrossRef]
21. Majmundar, A.J.; Wong, W.J.; Simon, M.C. Hypoxia-inducible factors and the response to hypoxic stress. *Mol. Cell* **2010**, *40*, 294–309. [CrossRef]
22. Courtney, K.D.; Ma, Y.; Diaz de Leon, A.; Christie, A.; Xie, Z.; Woolford, L.; Singla, N.; Joyce, A.; Hill, H.; Madhuranthakam, A.J.; et al. HIF-2 Complex Dissociation, Target Inhibition, and Acquired Resistance with PT2385, a First-in-Class HIF-2 Inhibitor, in Patients with Clear Cell Renal Cell Carcinoma. *Clin. Cancer Res.* **2020**, *26*, 793–803. [CrossRef]
23. Stransky, L.A.; Vigeant, S.M.; Huang, B.; West, D.; Denize, T.; Walton, E.; Signoretti, S.; Kaelin, W.G., Jr. Sensitivity of VHL mutant kidney cancers to HIF2 inhibitors does not require an intact p53 pathway. *Proc. Natl. Acad. Sci. USA* **2022**, *119*, e2120403119. [CrossRef]
24. Chen, Y.; Cattoglio, C.; Dailey, G.M.; Zhu, Q.; Tjian, R.; Darzacq, X. Mechanisms governing target search and binding dynamics of hypoxia-inducible factors. *eLife* **2022**, *11*, e75064. [CrossRef]
25. Choueiri, T.K.; Hessel, C.; Halabi, S.; Sanford, B.; Michaelson, M.D.; Hahn, O.; Walsh, M.; Olencki, T.; Picus, J.; Small, E.J.; et al. Cabozantinib versus sunitinib as initial therapy for metastatic renal cell carcinoma of intermediate or poor risk (Alliance A031203 CABOSUN randomised trial): Progression-free survival by independent review and overall survival update. *Eur. J. Cancer* **2018**, *94*, 115–125. [CrossRef]
26. Motzer, R.J.; Powles, T.; Burotto, M.; Escudier, B.; Bourlon, M.T.; Shah, A.Y.; Suarez, C.; Hamzaj, A.; Porta, C.; Hocking, C.M.; et al. Nivolumab plus cabozantinib versus sunitinib in first-line treatment for advanced renal cell carcinoma (CheckMate 9ER): Long-term follow-up results from an open-label, randomised, phase 3 trial. *Lancet Oncol.* **2022**, *23*, 888–898. [CrossRef]
27. Organ, S.L.; Tsao, M.S. An overview of the c-MET signaling pathway. *Ther. Adv. Med. Oncol.* **2011**, *3*, S7–S19. [CrossRef]
28. Oh, R.R.; Park, J.Y.; Lee, J.H.; Shin, M.S.; Kim, H.S.; Lee, S.K.; Kim, Y.S.; Lee, S.H.; Lee, S.N.; Yang, Y.M.; et al. Expression of HGF/SF and Met protein is associated with genetic alterations of VHL gene in primary renal cell carcinomas. *APMIS* **2002**, *110*, 229–238. [CrossRef] [PubMed]
29. Choueiri, T.K.; McDermott, D.F.; Merchan, J.; Bauer, T.M.; Figlin, R.; Heath, E.I.; Michaelson, M.D.; Arrowsmith, E.; D'Souza, A.; Zhao, S.; et al. Belzutifan plus cabozantinib for patients with advanced clear cell renal cell carcinoma previously treated with immunotherapy: An open-label, single-arm, phase 2 study. *Lancet Oncol.* **2023**, *24*, 553–562. [CrossRef] [PubMed]
30. Choueiri, T.K.; Bauer, T.; Merchan, J.R.; McDermott, D.F.; Figlin, R.; Arrowsmith, E.; Michaelson, M.D.; Heath, E.; D'Souza, A.A.; Zhao, S.; et al. LBA87 Phase II LITESPARK-003 study of belzutifan in combination with cabozantinib for advanced clear cell renal cell carcinoma (ccRCC). *Ann. Oncol.* **2023**, *34*, S1328–S1329. [CrossRef]
31. Capozzi, M.; De Divitiis, C.; Ottaiano, A.; von Arx, C.; Scala, S.; Tatangelo, F.; Delrio, P.; Tafuto, S. Lenvatinib, a molecule with versatile application: From preclinical evidence to future development in anti-cancer treatment. *Cancer Manag. Res.* **2019**, *11*, 3847–3860. [CrossRef]
32. Choueiri, T.K.; Eto, M.; Motzer, R.; De Giorgi, U.; Buchler, T.; Basappa, N.S.; Mendez-Vidal, M.J.; Tjulandin, S.; Hoon Park, S.; Melichar, B.; et al. Lenvatinib plus pembrolizumab versus sunitinib as first-line treatment of patients with advanced renal cell carcinoma (CLEAR): Extended follow-up from the phase 3, randomised, open-label study. *Lancet Oncol.* **2023**, *24*, 228–238. [CrossRef] [PubMed]
33. Motzer, R.J.; Hutson, T.E.; Glen, H.; Michaelson, M.D.; Molina, A.; Eisen, T.; Jassem, J.; Zolnierek, J.; Maroto, J.P.; Mellado, B.; et al. Lenvatinib, everolimus, and the combination in patients with metastatic renal cell carcinoma: A randomised, phase 2, open-label, multicentre trial. *Lancet Oncol.* **2015**, *16*, 1473–1482. [CrossRef]
34. Albiges, L.; Beckermann, K.; Miller, W.H.; Goh, J.C.; Gajate, P.; Harris, C.A.; Suarez, C.; Peer, A.; Park, S.H.; Stadler, W.M.; et al. Belzutifan plus lenvatinib for patients (pts) with advanced clear cell renal cell carcinoma (ccRCC) after progression on a PD-1/L1 and VEGF inhibitor: Preliminary results of arm B5 of the phase 1/2 KEYMAKER-U03B study. *J. Clin. Oncol.* **2023**, *41*, 4553. [CrossRef]
35. Motzer, R.J.; Schmidinger, M.; Eto, M.; Suarez, C.; Figlin, R.; Liu, Y.; Perini, R.; Zhang, Y.; Heng, D.Y. LITESPARK-011: Belzutifan plus lenvatinib vs cabozantinib in advanced renal cell carcinoma after anti-PD-1/PD-L1 therapy. *Future Oncol.* **2023**, *19*, 113–121. [CrossRef]
36. Bindra, R.S.; Vasselli, J.R.; Stearman, R.; Linehan, W.M.; Klausner, R.D. VHL-mediated hypoxia regulation of cyclin D1 in renal carcinoma cells. *Cancer Res.* **2002**, *62*, 3014–3019.
37. VanArsdale, T.; Boshoff, C.; Arndt, K.T.; Abraham, R.T. Molecular Pathways: Targeting the Cyclin D-CDK4/6 Axis for Cancer Treatment. *Clin. Cancer Res.* **2015**, *21*, 2905–2910. [CrossRef] [PubMed]
38. Nicholson, H.E.; Tariq, Z.; Housden, B.E.; Jennings, R.B.; Stransky, L.A.; Perrimon, N.; Signoretti, S.; Kaelin, W.G., Jr. HIF-independent synthetic lethality between CDK4/6 inhibition and VHL loss across species. *Sci. Signal.* **2019**, *12*, eaay0482. [CrossRef] [PubMed]

39. Cristofanilli, M.; Turner, N.C.; Bondarenko, I.; Ro, J.; Im, S.A.; Masuda, N.; Colleoni, M.; DeMichele, A.; Loi, S.; Verma, S.; et al. Fulvestrant plus palbociclib versus fulvestrant plus placebo for treatment of hormone-receptor-positive, HER2-negative metastatic breast cancer that progressed on previous endocrine therapy (PALOMA-3): Final analysis of the multicentre, double-blind, phase 3 randomised controlled trial. *Lancet Oncol.* **2016**, *17*, 425–439. [CrossRef] [PubMed]
40. Choueiri, T.K.; Powles, T.; Albiges, L.; Burotto, M.; Szczylik, C.; Zurawski, B.; Yanez Ruiz, E.; Maruzzo, M.; Suarez Zaizar, A.; Fein, L.E.; et al. Cabozantinib plus Nivolumab and Ipilimumab in Renal-Cell Carcinoma. *N. Engl. J. Med.* **2023**, *388*, 1767–1778. [CrossRef]
41. Choueiri, T.K.; Plimack, E.R.; Powles, T.; Voss, M.H.; Gurney, H.; Silverman, R.K.; Perini, R.F.; Rodriguez-Lopez, K.; Rini, B.I. Phase 3 study of first-line treatment with pembrolizumab + belzutifan + lenvatinib or pembrolizumab/quavonlimab + lenvatinib versus pembrolizumab + lenvatinib for advanced renal cell carcinoma (RCC). *J. Clin. Oncol.* **2022**, *40*, TPS399. [CrossRef]
42. Ravaud, A.; Motzer, R.J.; Pandha, H.S.; George, D.J.; Pantuck, A.J.; Patel, A.; Chang, Y.H.; Escudier, B.; Donskov, F.; Magheli, A.; et al. Adjuvant Sunitinib in High-Risk Renal-Cell Carcinoma after Nephrectomy. *N. Engl. J. Med.* **2016**, *375*, 2246–2254. [CrossRef] [PubMed]
43. Haas, N.B.; Manola, J.; Dutcher, J.P.; Flaherty, K.T.; Uzzo, R.G.; Atkins, M.B.; DiPaola, R.S.; Choueiri, T.K. Adjuvant Treatment for High-Risk Clear Cell Renal Cancer: Updated Results of a High-Risk Subset of the ASSURE Randomized Trial. *JAMA Oncol.* **2017**, *3*, 1249–1252. [CrossRef] [PubMed]
44. Choueiri, T.K.; Tomczak, P.; Park, S.H.; Venugopal, B.; Ferguson, T.; Chang, Y.H.; Hajek, J.; Symeonides, S.N.; Lee, J.L.; Sarwar, N.; et al. Adjuvant Pembrolizumab after Nephrectomy in Renal-Cell Carcinoma. *N. Engl. J. Med.* **2021**, *385*, 683–694. [CrossRef]
45. Motzer, R.J.; Russo, P.; Grunwald, V.; Tomita, Y.; Zurawski, B.; Parikh, O.; Buti, S.; Barthelemy, P.; Goh, J.C.; Ye, D.; et al. Adjuvant nivolumab plus ipilimumab versus placebo for localised renal cell carcinoma after nephrectomy (CheckMate 914): A double-blind, randomised, phase 3 trial. *Lancet* **2023**, *401*, 821–832. [CrossRef]
46. Pal, S.K.; Uzzo, R.; Karam, J.A.; Master, V.A.; Donskov, F.; Suarez, C.; Albiges, L.; Rini, B.; Tomita, Y.; Kann, A.G.; et al. Adjuvant atezolizumab versus placebo for patients with renal cell carcinoma at increased risk of recurrence following resection (IMmotion010): A multicentre, randomised, double-blind, phase 3 trial. *Lancet* **2022**, *400*, 1103–1116. [CrossRef]
47. Choueiri, T.K.; Bedke, J.; Karam, J.A.; McKay, R.R.; Motzer, R.J.; Pal, S.K.; Suarez, C.; Uzzo, R.; Liu, H.; Burgents, J.E.; et al. LITESPARK-022: A phase 3 study of pembrolizumab + belzutifan as adjuvant treatment of clear cell renal cell carcinoma (ccRCC). *J. Clin. Oncol.* **2022**, *40*, TPS4602. [CrossRef]
48. Chauvin, J.M.; Zarour, H.M. TIGIT in cancer immunotherapy. *J. Immunother. Cancer* **2020**, *8*, e000957. [CrossRef]
49. Kim, T.W.; Bedard, P.L.; LoRusso, P.; Gordon, M.S.; Bendell, J.; Oh, D.Y.; Ahn, M.J.; Garralda, E.; D'Angelo, S.P.; Desai, J.; et al. Anti-TIGIT Antibody Tiragolumab Alone or with Atezolizumab in Patients With Advanced Solid Tumors: A Phase 1a/1b Nonrandomized Controlled Trial. *JAMA Oncol.* **2023**, *9*, 1574–1582. [CrossRef]
50. Cho, B.C.; Abreu, D.R.; Hussein, M.; Cobo, M.; Patel, A.J.; Secen, N.; Lee, K.H.; Massuti, B.; Hiret, S.; Yang, J.C.H.; et al. Tiragolumab plus atezolizumab versus placebo plus atezolizumab as a first-line treatment for PD-L1-selected non-small-cell lung cancer (CITYSCAPE): Primary and follow-up analyses of a randomised, double-blind, phase 2 study. *Lancet Oncol.* **2022**, *23*, 781–792. [CrossRef]
51. Mettu, N.B.; Ulahannan, S.V.; Bendell, J.C.; Garrido-Laguna, I.; Strickler, J.H.; Moore, K.N.; Stagg, R.; Kapoun, A.M.; Faoro, L.; Sharma, S. A Phase 1a/b Open-Label, Dose-Escalation Study of Etigilimab Alone or in Combination with Nivolumab in Patients with Locally Advanced or Metastatic Solid Tumors. *Clin. Cancer Res.* **2022**, *28*, 882–892. [CrossRef] [PubMed]
52. Niu, J.; Maurice-Dror, C.; Lee, D.H.; Kim, D.W.; Nagrial, A.; Voskoboynik, M.; Chung, H.C.; Mileham, K.; Vaishampayan, U.; Rasco, D.; et al. First-in-human phase 1 study of the anti-TIGIT antibody vibostolimab as monotherapy or with pembrolizumab for advanced solid tumors, including non-small-cell lung cancer. *Ann. Oncol.* **2022**, *33*, 169–180. [CrossRef]
53. Toledo, R.A.; Jimenez, C.; Armaiz-Pena, G.; Arenillas, C.; Capdevila, J.; Dahia, P.L.M. Hypoxia-Inducible Factor 2 Alpha (HIF2alpha) Inhibitors: Targeting Genetically Driven Tumor Hypoxia. *Endocr. Rev.* **2023**, *44*, 312–322. [CrossRef]
54. Ma, Y.; Joyce, A.; Brandenburg, O.; Saatchi, F.; Stevens, C.; Toffessi Tcheuyap, V.; Christie, A.; Do, Q.N.; Fatunde, O.; Macchiaroli, A.; et al. HIF2 Inactivation and Tumor Suppression with a Tumor-Directed RNA-Silencing Drug in Mice and Humans. *Clin. Cancer Res.* **2022**, *28*, 5405–5418. [CrossRef]
55. Brugarolas, J.; Beckermann, K.; Rini, B.I.; Vogelzang, N.J.; Lam, E.T.; Hamilton, J.C.; Schluep, T.; Yi, M.; Wong, S.; Gamelin, E.; et al. Initial results from the phase 1 study of ARO-HIF2 to silence HIF2-alpha in patients with advanced ccRCC (AROHIF21001). *J. Clin. Oncol.* **2022**, *40*, 339. [CrossRef]
56. Lu, J.; Wei, H.; Sun, W.; Geng, J.; Liu, K.; Liu, J.; Liu, Z.; Fu, J.; He, Y.; Wang, K.; et al. Abstract 6330: NKT2152: A highly potent HIF2$\alpha$ inhibitor and its therapeutic potential in solid tumors beyond ccRCC. *Cancer Res.* **2022**, *82*, 6330. [CrossRef]
57. Lawson, K.V.; Sivick Gauthier, K.E.; Mailyan, A.K.; Fournier, J.T.; Beatty, J.W.; Drew, S.L.; Kalisiak, J.; Gal, B.; Mata, G.; Wang, Z.; et al. Abstract 1206: Discovery and characterization of AB521, a novel, potent, and selective hypoxia-inducible factor (HIF)-2$\alpha$ inhibitor. *Cancer Res.* **2021**, *81*, 1206. [CrossRef]
58. Sivick Gauthier, K.E.; Piovesan, D.; Cho, S.; Lawson, K.V.; Schweickert, P.G.; Lopez, A.; Liu, S.; Park, T.; Mailyan, A.; Fournier, J.T.A.; et al. Abstract P206: AB521 potently and selectively inhibits pro-tumorigenic gene transcription by Hypoxia-Inducible Factor (HIF)-2$\alpha$ in vitro and in vivo. *Cancer Res.* **2021**, *20*, P206. [CrossRef]

59. Lawson, K.V.; Sivick Gauthier, K.E.; Piovesan, D.; Mailyan, A.; Mata, G.; Fournier, J.T.; Yu, K.; Liu, S.; Soriano, F.; Jin, L.; et al. 46P AB521, a clinical-stage, potent, and selective Hypoxia-Inducible Factor (HIF)-2α inhibitor, for the treatment of renal cell carcinoma. *Ann. Oncol.* **2022**, *33*, S21. [CrossRef]
60. Liao, K.; Foster, P.; Seitz, L.; Cheng, T.; Gauthier, K.; Lawson, K.; Jin, L.; Paterson, E. HIF-2α inhibitor AB521 modulates erythropoietin levels in healthy volunteers following a single oral dose. *Eur. J. Cancer* **2022**, *174*, S20. [CrossRef]
61. Wang, P.; Yang, R.; Sun, Y.; Ju, X.; Zhao, J.; Liu, Y.; Liu, X.; Zou, Z.; Ren, J.; Wang, M.; et al. Abstract 494: BPI-452080: A potent and selective HIF-2α inhibitor for the treatment of clear cell renal cell carcinoma, von Hippel-Lindau disease, and other solid tumors. *Cancer Res.* **2023**, *83*, 494. [CrossRef]
62. Yang, L.; Fan, Y.; Zhang, Q. Targeting ferroptosis in renal cell carcinoma: Potential mechanisms and novel therapeutics. *Heliyon* **2023**, *9*, e18504. [CrossRef]
63. Koh, M.Y. The identification of a novel orally available ferroptosis inducer for the treatment of clear cell renal carcinoma. *Oncologist* **2023**, *28*, S10. [CrossRef]
64. Bai, J.; Chen, W.B.; Zhang, X.Y.; Kang, X.N.; Jin, L.J.; Zhang, H.; Wang, Z.Y. HIF-2alpha regulates CD44 to promote cancer stem cell activation in triple-negative breast cancer via PI3K/AKT/mTOR signaling. *World J. Stem Cells* **2020**, *12*, 87–99. [CrossRef]
65. Singhal, R.; Mitta, S.R.; Das, N.K.; Kerk, S.A.; Sajjakulnukit, P.; Solanki, S.; Andren, A.; Kumar, R.; Olive, K.P.; Banerjee, R.; et al. HIF-2alpha activation potentiates oxidative cell death in colorectal cancers by increasing cellular iron. *J. Clin. Investig.* **2021**, *131*, e143691. [CrossRef]
66. Zhao, J.; Du, F.; Shen, G.; Zheng, F.; Xu, B. The role of hypoxia-inducible factor-2 in digestive system cancers. *Cell Death Dis.* **2015**, *6*, e1600. [CrossRef] [PubMed]
67. Pavlakis, D.; Kampantais, S.; Gkagkalidis, K.; Gourvas, V.; Memmos, D.; Tsionga, A.; Dimitriadis, G.; Vakalopoulos, I. Hypoxia-Inducible Factor 2a Expression Is Positively Correlated With Gleason Score in Prostate Cancer. *Technol. Cancer Res. Treat.* **2021**, *20*, 1533033821990010. [CrossRef] [PubMed]
68. Buscheck, F.; Fraune, C.; Simon, R.; Kluth, M.; Hube-Magg, C.; Moller-Koop, C.; Sarper, I.; Ketterer, K.; Henke, T.; Eichelberg, C.; et al. Prevalence and clinical significance of VHL mutations and 3p25 deletions in renal tumor subtypes. *Oncotarget* **2020**, *11*, 237–249. [CrossRef] [PubMed]

cancers

MDPI

*Article*

# Association Between CKAP4 Expression and Poor Prognosis in Patients with Bladder Cancer Treated with Radical Cystectomy

**Hiroki Katsumata [1], Dai Koguchi [1], Shuhei Hirano [1], Anna Suzuki [2], Kengo Yanagita [3], Yuriko Shimizu [1], Wakana Hirono [4], Soichiro Shimura [1], Masaomi Ikeda [1], Hideyasu Tsumura [1], Daisuke Ishii [1], Yuichi Sato [1,5] and Kazumasa Matsumoto [1,*]**

[1] Department of Urology, Kitasato University School of Medicine, 1-15-1 Kitasato Minami-ku, Sagamihara 252-0374, Japan; giri_giri_zin_zin@yahoo.co.jp (H.K.)
[2] Department of Pathology, Nagaoka Chuo General Hospital, 2041 Kawasaki, Nagaoka 940-0861, Japan
[3] Biofluid Biomarker Center, Niigata University, 8050 ikarashi 2-no-cho Nishi-ku, Niigata 950-2181, Japan
[4] Kitasato University School of Medicine, 1-15-1 Kitasato Minami-ku, Sagamihara 252-0374, Japan
[5] KITASATO-OTSUKA Biomedical Assay Laboratories Co., Ltd., 1-15-1 Kitasato, Minami-ku, Sagamihara 252-0329, Japan
* Correspondence: kazumasa@cd5.so-net.ne.jp; Tel.: +81-42-778-9091; Fax: +81-42-778-9374

**Simple Summary:** Cytoskeleton-associated protein 4 (CKAP4), which has been linked to worse outcomes in several types of cancer, has emerged as a novel biomarker to predict patient outcomes for bladder cancer following radical cystectomy. This study investigated CKAP4 levels in bladder cancer specimens after radical cystectomy, and the association between CKAP4 levels, clinicopathological characteristics, and patient outcomes was analyzed. The analysis revealed that CKAP4 was connected to a higher risk of cancer recurrence, which means that CKAP4 could be a useful clinical tool to predict cancer recurrence after surgery.

**Abstract: Background/Objectives**: While cytoskeleton-associated protein 4 (CKAP4) has been associated with prognosis in various malignancies, its prognostic value for bladder cancer (BCa) remains unclear. The aim of this study was to evaluate CKAP4 expression in tumor cells and cancer-associated fibroblasts (CAFs) following radical cystectomy (RC) in patients with BCa. **Methods**: In this study, CKAP4 in tumor cells was defined as CKAP4-1, while CKAP4 expressed in CAFs was defined as CKAP4-2. CKAP4-2 expression was evaluated to explore its potential association with tumor aggressiveness and patient outcomes. CKAP4 expression in 86 RC specimens was assessed using immunohistochemistry. CKAP4-1 positivity was considered when $\geq$5% cytoplasmic staining of cancer cells, with at least moderate staining intensity, was observed. CKAP4-2 positivity was evaluated using a point scale (0–3), with scores based on the number of CKAP4 positive CAFs in the tumor stroma. Scores of 2 (moderate number of CAFs) and 3 (significant number of CAFs) were considered to indicate positivity. **Results**: CKAP4-1 and CKAP4-2 were expressed in 53 (61.6%) and 34 (39.5%) patients, respectively. Kaplan–Meier analysis showed that patients with CKAP4-1 had significantly shorter cancer-specific survival and recurrence-free survival (RFS; $p = 0.046$ and $p = 0.0173$, respectively). Multivariate analysis showed that CKAP4-1 positivity was an independent predictor of RFS ($p = 0.041$, hazard ratio: 2.09, 95% confidence interval: 1.03–4.25). **Conclusions**: This study showed that CKAP4 expression in tumor cells may serve as a useful prognostic biomarker for patients with BCa who undergo RC.

 https://doi.org/10.3390/cancers17081278

**Keywords:** bladder cancer; radical cystectomy; CKAP4; cancer-associated fibroblasts; immunohistochemistry; prognosis

## 1. Introduction

Bladder cancer (BCa) is the tenth most prevalent cancer globally [1]. Radical cystectomy (RC) with bilateral pelvic lymph node (LN) dissection remains the gold standard treatment for patients with muscle-invasive BCa (MIBC) and non-MIBC (NMIBC) refractory to bacillus Calmette-Guérin. Although the surgical technique has advanced and the role of pelvic lymphadenectomy is better understood, the 5-year overall survival rate after RC without neoadjuvant chemotherapy (NAC) or adjuvant chemotherapy (AC) is approximately 50% [2]. Several clinicopathological factors, such as tumor grade, lymphovascular invasion (LVI), and LN status have been studied in an effort to improve unfavorable prognoses after RC [3]. These risk factors facilitate estimation of the recurrence risk and survival outcomes; however, they do not predict individual patient prognosis. Some studies reported the efficacy of new prognostic biomarkers for BCa after RC, but none of them are available for clinical use [4–6]; therefore, novel prognostic biomarkers for BCa are necessary.

Cytoskeleton-associated protein 4 (CKAP4), also termed CLIMP-63 or p63, is a type II transmembrane protein with reversible palmitoylation [7,8]. It is mainly found in the rough endoplasmic reticulum, where it stabilizes its structure [9]. CKAP4 has been associated with various cancers and has attracted considerable attention in recent years. In hepatocellular carcinoma, esophageal cancer, and renal cell carcinoma, CKAP4 overexpression was significantly associated with unfavorable outcomes [10–12]. CKAP4 in pancreatic cancer has emerged as a potential therapeutic target for the inhibition of DKK1-CKAP4 binding and Akt activation [13]. Although the literature on CKAP4 remains scarce, a basic study has intensively investigated the mechanism of CKAP4 in the progression of BCa, indicating its potential as a novel biomarker [14]. Verification of the prognostic value of CKAP4 in clinical practice is essential for better BCa management because patients who require RC with curative intent often have poor outcomes.

The aims of this study were to evaluate the expression levels of CKAP4 in tumor cells and cancer-associated fibroblasts (CAFs) using immunohistochemical analyses of archived RC specimens and to assess the impact of CKAP4 on the prognosis of patients who underwent RC for BCa. In this study, to investigate their associations with clinicopathological features and patient outcomes, CKAP4 expressed in tumor cells was defined as CKAP4-1, while CKAP4 expressed in CAFs was defined as CKAP4-2.

## 2. Materials and Methods

### 2.1. Patients

We retrospectively analyzed the clinical data and archived specimens of patients with BCa who underwent RC with pelvic and iliac lymphadenectomies between 1990 and 2015 at the Kitasato University Hospital (Kanagawa, Japan). We also examined normal urothelial tissue specimens from adjacent tumor tissues using NMIBCs as a negative control. RC was performed in patients with pathologically proven MIBC and in those with NMIBC who failed to respond to intravesical therapy [15]. Patient characteristics were obtained from medical records, including age at RC, sex, pathological status (pT and pN stages), tumor grade, LVI status, carcinoma in situ (CIS), history of AC and salvage chemotherapy (SC), recurrence, and cancer-specific death. Pathological staging was performed according to the 2002 TNM classification. Pathological grading was performed according to the 1973 World Health Organization classification. LVI was defined as the presence of cancer cells within the endothelial space; however, cancer cells merely invading the vascular lumen were considered negative [16]. AC was performed in patients with pT $\geq$ 3 disease or a positive LN status. All the patients with AC or SC received platinum-based chemotherapy. Because this was a retrospective study, all patients found to be eligible during the study period were included, and no formal sample size calculation was performed. This study was conducted in accordance with the guidelines of the Declaration of Helsinki and was approved by the Ethics Committee of Kitasato University School of Medicine on 24 May 2017 (B17-010). Participants were informed of the study and provided with the opportunity to opt out at any time. As this was a retrospective study, individual written informed consent was not required.

### 2.2. Immunohistochemistry and Scoring

Three-micrometer thick sections from 10% formalin-fixed and paraffin-embedded BCa tissue blocks were deparaffinized in xylene and rehydrated in a descending ethanol series. After treatment with 3% hydrogen peroxide for 10 min, the antigen was retrieved by autoclaving in Tris-EDTA buffer (0.01 M Tris-hydroxymethyl aminomethane, 0.001 M EDTA-2Na, pH 9.0) at 121 °C for 10 min. After washing in Tris-buffered saline (TBS) for 5 min and blocking in 0.5% casein for 10 min, the sections were reacted with 1000 times diluted anti-CKAP4 polyclonal antibody (HPA000792; Sigma Life Science, St. Louis, MO, USA) for 18 h at room temperature. After rinsing with TBS three times for 5 min each, samples were treated with horseradish peroxidase-labeled polymer reagent (EnVisoin+ Dual Link System-HRP Kit; Dako, Glostrup, Denmark) for 30 min at room temperature. After rinsing with TBS three times for 5 min each, the sections were visualized with a Stable DAB solution (Invitrogen, Carlsbad, CA, USA) and counterstained with Mayer's hematoxylin. The protocol for immunohistochemical analyses was based on a previously reported method by Yanagita et al. [17].

The expression levels of CKAP4 were evaluated in tumor cells and in CAFs located in the tumor stroma. We defined CKAP4-1 as CKAP4 in tumor cells and CKAP4-2 as CKAP4 in CAFs. CKAP4-1 was immunohistochemically evaluated by determining the intensity and percentage of positive tumor cells. Normal urothelial cells were used as an internal control. The staining intensity of the tumor cell cytoplasm was categorized into four groups: 0, no staining; 1, weak staining; 2, moderate staining; and 3, strong staining. CKAP4-1 positivity was defined as cases where $\geq$5% of tumor cells exhibited a staining score of 2 or 3. These scoring criteria were adapted from the study by Nagoya et al. [18]. Immunohistochemical staining for CKAP4-2 was performed based on the number of CKAP4 positive CAFs in the tumor stroma. We categorized CKAP4-2 expression into four groups: 0, no staining of CAF; 1, few CAFs; 2, moderate number of CAFs; and 3, significant number of CAFs. CKAP4-2 positivity was defined as patients with scores of 2 or 3. Because no study has evaluated CKAP4 expression in CAFs, this scoring method was adapted from a previous report by Akanda et al. [19]. Two investigators (H.K. and Y.S.) blinded to the clinical and pathological data, reviewed all the immunostained sections. Discordant cases were reviewed and discussed until consensus was reached.

### 2.3. Statistical Analyses

In immunohistochemical analysis, the age (<65 vs. $\geq$65), pathological stage (pT $\leq$ 2 vs. $\geq$3), LN status (N0 vs. N1 and N2), and pathological grade (1 and 2 vs. 3) were evaluated as dichotomized variables. The association between CKAP4 expression and clinicopathological status (sex, age, pathological stage, LN status, pathological grade, LVI, CIS, history of AC and SC, recurrence, and cancer-specific death) was evaluated using Fisher's exact test. The correlation between CKAP4-1 and CKAP4-2 expression was also examined. Cancer-specific survival (CSS) and recurrence-free survival (RFS) were estimated using the Kaplan–Meier method with log-rank tests. Univariate and multivariate analyses were performed using Cox proportional hazards analysis to estimate the association between CKAP4 expression and clinicopathological variables. Statistical significance was set at $p < 0.05$. All reported $p$ values are two-sided. Stata 17 for Windows (Stata, Chicago, IL, USA) was used for all analyses.

## 3. Results

The clinical data and archived specimens of 125 patients with BCa who underwent RC with pelvic and iliac lymphadenectomies were initially included. Of the total, 39 patients were excluded for the following reasons: 10 with histological variants, including squamous cell carcinoma, adenocarcinoma, and small cell carcinoma; 15 who had been previously treated with NAC; and 14 who were lost to follow-up. The final study group (n = 86) comprised 67 men (78%) and 19 women (22%). None of the patients received preoperative NAC or radiotherapy, and no distant metastases were observed at the time of diagnosis.

### 3.1. Immunohistochemistry

Figure 1 shows CKAP4 staining in normal urothelial and tumor tissues of the study group. In normal urothelial tissues, umbrella cells showed strong cytoplasmic staining, whereas the normal urothelium showed negative to weak staining, with scattered weakly positive normal fibroblasts (Figure 1a). In tumor tissues, CKAP4 was observed in the cytoplasm of tumor cells and CAFs at various degrees and intensities (Figure 1b–d).

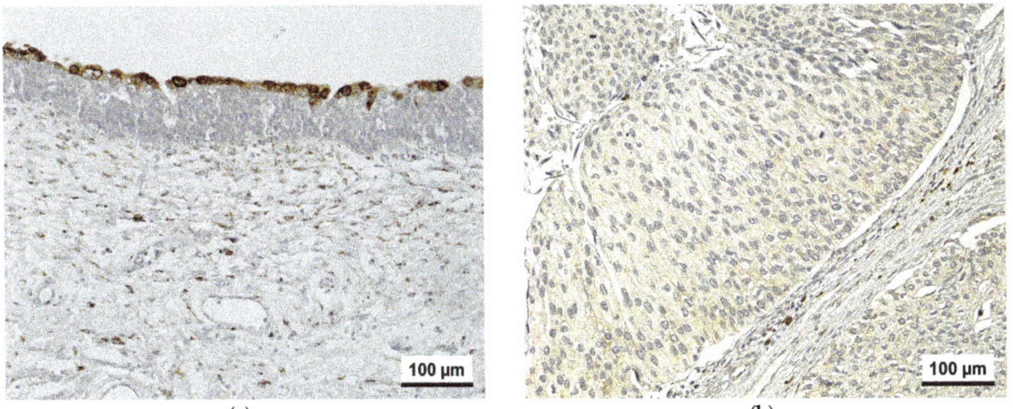


(a) (b)

**Figure 1.** *Cont.*

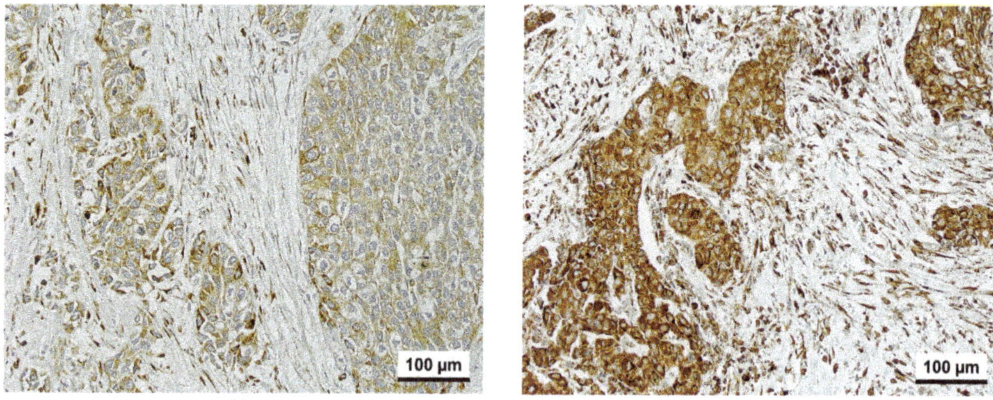


(c) (d)

**Figure 1.** CKAP4 immunoreactivity in normal urothelium and bladder cancer tissue (200× magnification). (**a**) Normal urothelium. (**b**) Tumor cells with weak staining and a few positive CAFs. (**c**) Tumor cells with moderate staining and a moderate number of positive CAFs. (**d**) Tumor cells with strong staining and a significant number of positive CAFs. CKAP4, cytoskeleton-associated protein 4; CAFs, cancer-associated fibroblasts.

### *3.2. Association of CKAP4 Expression with Clinicopathological Characteristics*

Table 1 summarizes the clinicopathological characteristics of the 86 patients. CKAP4-1 and CKAP4-2 were positive in 61.6% (n = 53) and 39.5% (n = 34) of the patients, respectively. During a median follow-up of 36.0 months, the proportion of patients who died of BCa or other causes was 44.2% (n = 38) and 14.0% (n = 12), respectively. The CKAP4-1 positive group had significantly higher proportions of patients with pT $\geq$ 3 (67.9% vs. 39.4%) and positive LVI (72.0% vs. 48.4%) compared to the CKAP4-1 negative group. Similarly, the CKAP4-2 positive group had significantly higher proportions of patients with pT $\geq$ 3 (73.5% vs. 46.2%), positive LVI (84.4% vs. 49.0%), positive LN metastasis (40.6% vs. 16.0%), Grade 3 tumors (82.4% vs. 46.2%), and a history of AC (44.1% vs. 11.5%) than did the CKAP4-2 negative group.

**Table 1.** Relationship between CKAP4 and clinicopathological characteristics.

| Characteristics | | CKAP4-1 | | | CKAP4-2 | | |
|---|---|---|---|---|---|---|---|
| | Total No. (%) | Negative (%) | Positive (%) | *p*-Value | Negative (%) | Positive (%) | *p*-Value |
| Overall | 86 | 33 (38.4) | 53 (61.6) | | 52 (60.5) | 34 (39.5) | |
| Age, years | | | | | | | |
| Median (IQR) | 65 (57–71) | 65 (56–71) | 64 (57–72) | | 65 (57–70) | 64 (56–72) | |
| <65 | 42 (48.8) | 15 (45.5) | 27 (50.9) | 0.66 | 24 (46.2) | 18 (52.9) | 0.66 |
| ≥65 | 44 (51.2) | 18 (54.6) | 26 (49.1) | | 28 (53.9) | 16 (47.1) | |
| Sex | | | | | | | |
| Male | 67 (77.9) | 28 (84.9) | 39 (73.6) | 0.28 | 42 (80.8) | 25 (75.5) | 0.43 |
| Female | 19 (22.1) | 5 (15.2) | 14 (26.4) | | 10 (19.2) | 9(26.7) | |
| pT stage | | | | | | | |
| pT ≤ 2 | 37 (43.0) | 20 (60.6) | 17 (32.1) | 0.014 | 28 (53.9) | 9 (26.5) | 0.015 |
| pT ≥ 3 | 49 (57.0) | 13 (39.4) | 36 (67.9) | | 24 (46.2) | 25 (73.5) | |
| Lymph node status | | | | | | | |
| N0 | 61 (74.4) | 26 (83.9) | 35 (68.6) | 0.19 | 42 (84.0) | 19 (59.4) | 0.019 |
| N+ | 21 (25.6) | 5 (16.1) | 16 (31.4) | | 8 (16.0) | 13 (40.6) | |
| Pathological grade | | | | | | | |
| G1–2 | 34 (39.5) | 16 (48.5) | 18 (34.0) | 0.25 | 28 (53.9) | 6 (17.6) | 0.001 |
| G3 | 52 (60.5) | 17 (51.5) | 35 (66.0) | | 24 (46.2) | 28 (82.4) | |

**Table 1.** *Cont.*

| Characteristics | | CKAP4-1 | | | CKAP4-2 | | |
|---|---|---|---|---|---|---|---|
| | Total No. (%) | Negative (%) | Positive (%) | *p*-Value | Negative (%) | Positive (%) | *p*-Value |
| LVI | | | | | | | |
| Negative | 30 (37.0) | 16 (51.6) | 14 (28.0) | 0.037 | 25 (51.0) | 5 (15.6) | 0.002 |
| Positive | 51 (63.0) | 15 (48.4) | 36 (72.0) | | 24 (49.0) | 27 (84.4) | |
| Carcinoma in situ | | | | | | | |
| Negative | 78 (90.7) | 31 (93.9) | 47 (88.7) | 0.70 | 47 (90.4) | 31 (91.2) | 1.000 |
| Positive | 8 (9.3) | 2 (6.1) | 6 (11.3) | | 5 (9.6) | 3 (8.8) | |
| Adjuvant chemotherapy | | | | | | | |
| No | 65 (75.6) | 27 (81.8) | 38 (71.7) | 0.31 | 46 (88.5) | 19 (55.9) | 0.001 |
| Yes | 21 (24.4) | 6 (18.2) | 15 (28.3) | | 6 (11.5) | 15 (44.1) | |
| Salvage chemotherapy | | | | | | | |
| No | 62 (72.1) | 26 (78.8) | 36 (67.9) | 0.32 | 38 (73.1) | 24 (70.6) | 0.81 |
| Yes | 24 (27.9) | 7 (21.2) | 17 (32.1) | | 14 (26.9) | 10 (29.4) | |
| Recurrence | | | | | | | |
| No | 43 (50.0) | 21 (63.6) | 22 (41.5) | 0.075 | 29 (55.8) | 14 (41.2) | 0.27 |
| Yes | 43 (50.0) | 12 (36.4) | 31 (58.5) | | 23 (44.2) | 20 (58.8) | |
| Cancer-specific death | | | | | | | |
| No | 48 (55.8) | 22 (66.7) | 26 (49.1) | 0.12 | 32 (61.5) | 16 (47.1) | 0.26 |
| Yes | 38 (44.2) | 11 (33.3) | 27 (50.9) | | 20 (38.5) | 18 (52.9) | |
| CKAP4-2 | | | | | | | |
| Negative | 52 (60.5) | 24 (72.7) | 28 (52.8) | 0.075 | | | |
| Positive | 34 (39.5) | 9 (27.3) | 25 (47.2) | | | | |

No., number; CKAP4, cytoskeleton-associated protein 4; LVI, lymphovascular invasion; IQR, interquartile range.

### 3.3. Survival Outcomes and CKAP4 Expression

Kaplan–Meier analysis showed that patients with positive CKAP4-1 had significantly shorter CSS and RFS ($p = 0.046$, and $p = 0.017$, respectively; Figure 2) than patients with negative CKAP4-1. Similarly, patients with positive CKAP4-2 had shorter CSS and RFS than patients with negative CKAP4-2, but the differences were not significant ($p = 0.085$ and $p = 0.058$, respectively; Figure 3). The median times to cancer death and recurrence for patients with positive CKAP4-1 were 39.3 and 28.8 months, respectively. However, for patients with negative CKAP4-1, the median times to cancer death and recurrence were not reached.

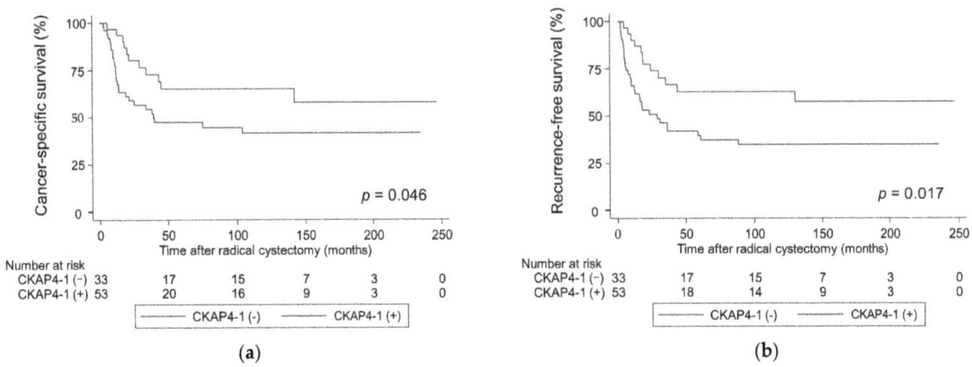


**Figure 2.** Probability of survival in patients with bladder cancer according to CKAP4-1 expression estimated using Kaplan–Meier analysis. (**a**) Cancer-specific survival; (**b**) Recurrence-free survival. CKAP4, cytoskeleton-associated protein 4.

In the univariate Cox regression analysis, independent risk factors for worse CSS and RFS were pT $\geq 3$, positive LN metastasis, and positive CKAP4-1. In the multivariate Cox regression analysis, positive CKAP4-1 was as an independent factor for worse RFS (hazard ratio: 2.09, 95% confidence interval: 1.03–4.25, $p$ = 0.041), whereas the other clinicopathological features did not have prognostic impact regarding CSS and RFS (Table 2).

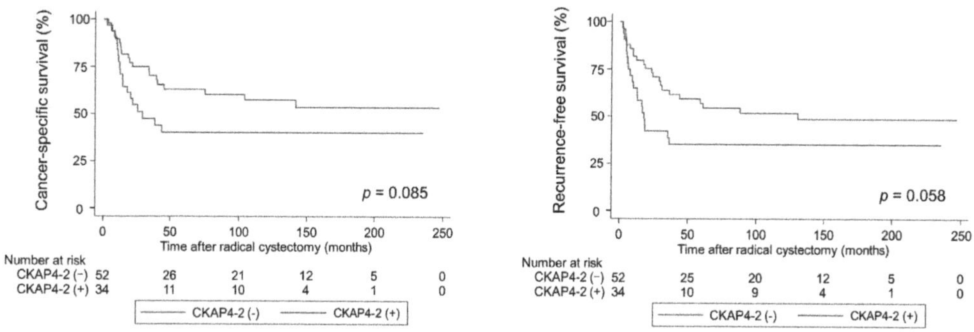


(**a**) (**b**)

**Figure 3.** Probability of survival in patients with bladder cancer according to CKAP4-2 expression estimated using Kaplan–Meier analysis. (**a**) Cancer-specific survival; (**b**) Recurrence-free survival. CKAP4, cytoskeleton-associated protein 4.

**Table 2.** Univariate and multivariate Cox proportional hazard analyses to predict cancer-specific survival and recurrence-free survival.

| | | **Cancer-Specific Survival** | | | | | |
|---|---|---|---|---|---|---|---|
| **Variables** | **Category** | **Univariate** | | | **Multivariate** | | |
| | | **HR** | **95% CI** | ***p*-Value** | **HR** | **95% CI** | ***p*-Value** |
| CKAP4-1 | Positive<br>Negative | 2.01<br>1 | 0.99–4.06 | 0.050 | 1.84<br>1 | 0.87–3.90 | 0.11 |
| CKAP4-2 | Positive<br>Negative | 1.74<br>1 | 0.91–3.30 | 0.089 | 1.24 | 0.56–2.73 | 0.58 |
| pT stage | $pT \geq 3$<br>$pT \leq 2$ | 2.25<br>1 | 1.13–4.47 | 0.021 | 1.78<br>1 | 0.80–3.93 | 0.15 |
| Pathological grade | G3<br>G1–2 | 1.31<br>1 | 0.67–2.57 | 0.42 | 0.92<br>1 | 0.43–1.99 | 0.84 |
| LN status | N+<br>N0 | 3.01<br>1 | 1.52–5.94 | 0.001 | 2.11<br>1 | 0.95–4.71 | 0.066 |
| CIS | Positive<br>Negative | 0.47<br>1 | 0.11–1.98 | 0.30 | 0.36<br>1 | 0.08–1.57 | 0.17 |
| | | **Recurrence-Free Survival** | | | | | |
| **Variables** | **Category** | **Univariate** | | | **Multivariate** | | |
| | | **HR** | **95% CI** | ***p*-Value** | **HR** | **95% CI** | ***p*-Value** |
| CKAP4-1 | Positive<br>Negative | 2.24<br>1 | 1.14–4.37 | 0.018 | 2.09<br>1 | 1.03–4.25 | 0.041 |
| CKAP4-2 | Positive<br>Negative | 1.77<br>1 | 0.97–3.24 | 0.062 | 1.32<br>1 | 0.62–2.79 | 0.45 |
| pT stage | $pT \geq 3$<br>$pT \leq 2$ | 2.11<br>1 | 1.11–4.08 | 0.021 | 1.75<br>1 | 0.85–3.61 | 0.12 |
| Pathological grade | G3<br>G1–2 | 1.23<br>1 | 0.65–2.30 | 0.51 | 0.84<br>1 | 0.41–1.72 | 0.64 |
| LN status | N+<br>N0 | 2.71<br>1 | 1.42–5.16 | 0.002 | 1.85<br>1 | 0.88–3.87 | 0.10 |
| CIS | Positive<br>Negative | 0.44<br>1 | 0.10–1.84 | 0.26 | 0.35<br>1 | 0.08–1.49 | 0.15 |

HR, hazard ratio; CI, confidence interval; CKAP4, cytoskeleton-associated protein 4; LN, lymph node; CIS, carcinoma in situ.

## 4. Discussion

In this retrospective analysis, we investigated the prognostic relevance of CKAP4 expression in patients with BCa who underwent RC and observed two striking findings. First, positive CKAP4 expression was associated with more aggressive pathological features, such as advanced pT stage and presence of LVI in tumor cells, and advanced pT stage, tumor grade 3, presence of LVI, and LN metastasis in CAFs. Second, the multivariate analysis adjusted for clinicopathological features revealed that CKAP4 expression in tumor cells was an independent prognostic factor for poor RFS, whereas CKAP4 expression in CAFs did not have a prognostic impact.

Although data on CKAP4 expression in BCa are scarce, pioneering work on the mechanism of CKAP4 expression in BCa cells has been reported by Sun et al. [14]. Biomarker analysis using the cell-SELEX method identified CKAP4 as having the highest affinity for the aptamer spl3, which had the best binding ability to BLCA 5637 cells. CKAP4 was involved in the progression of BCa through two potential mechanisms. First, CKAP4 facilitates cancer invasiveness by orchestrating a central-to-peripheral gradient of cell surface stiffness. For example, 5637 cells with CKAP4 exhibited a four-fold increase in motility compared to CKAP4-depleted 5637 cells, with the formation of lamellipodia in the CKAP4 positive cells presumably supporting the high migration potential. Second, exosomal CKAP4 promoted cancer metastasis. Injection of 5637 cells treated with CKAP4-containing exosome in mice exhibited approximately twice as much metastasis compared to the control 5637 cells. To assess the clinical relevance of CKAP4, the association between CKAP4 expression and the clinicopathological characteristics of patients with BCa was investigated by immunohistochemical analysis using a tissue microarray [14]. CKAP4 expression was observed in the tumor cytoplasm and nucleus and was significantly associated with MIBC and LN metastasis. As such, the mechanical properties of central-to-peripheral stiffness, lamellipodia formation, and exosomal CKAP4 significantly correlated with the invasiveness and metastasis of several cancers. In line with these findings, the worse prognostic value of CKAP4-1 demonstrated in the current multivariate analysis appears reasonable [18,20–22].

Another mechanism of CKAP4 underlying the biological aggressiveness of BCa is the interaction between CKAP4 and DKK1. The Wnt/β-catenin pathway plays a central role in cancer progression, and intensive research has found DKK1 as a Wnt/β-catenin pathway inhibitor [23,24]. However, recent preclinical and clinical studies have revealed that the interaction of DKK1 and CKAP4 on the cell membrane activates the Akt signaling pathway, wherein various types of cancers predominantly rely on cell proliferation [25,26]. In surgically resected esophageal squamous cell carcinoma, patients with both CKAP4 and DKK1 positivity showed significantly worse overall survival and RFS than those negative for both biomarkers [25]. Moreover, either DKK1 or CKAP4 knockdown suppressed cell growth in vitro and xenograft tumor formation via the Akt signaling pathway, and anti-CKAP4 antibody has also been demonstrated to show an antitumor effect in pancreatic cancer cells [25,26]. Although no study has reported an association between CKAP4 and DKK1, Shen et al., using western blotting, reported higher DKK1 expression levels in urothelial carcinoma tissues than in the normal urothelium [27]. The correlation between CKAP4 expression in tumor cells and poor RFS with advanced clinicopathological characteristics in the present study may represent the interaction of CKAP4 and DKK1. We believe that CKAP4 is worth highlighting as a novel prognostic and therapeutic biomarker for advanced BCa, and our findings encourage further studies to validate this hypothesis.

We observed varying levels of CKAP4 expression in CAFs within the tumor stroma, which were significantly associated with more aggressive clinicopathological features. These findings suggest that CKAP4 expression in CAFs may also be a useful biomarker for predicting the aggressive BCa phenotype. Although there are no reports on CKAP4 expression in CAFs, Gladka et al. identified CKAP4 as a novel biomarker for activated cardiac fibroblasts in a single-cell sequencing analysis, and CKAP4 was shown to be involved in myofibroblast activation in in vitro experiments [28]. Yang et al. reported the classification of CAFs into distinct subtypes that function as specific markers; in particular, myofibroblast-like CAFs (myCAFs) were found to be activated through direct interaction with cancer cells and exhibited dual tumor-restraining and tumor-promoting roles [29]. Du et al. performed an analysis using the TCGA database and reported that high expression of the myCAF marker gene was significantly associated with more advanced T-stage and worse overall survival and RFS in BCa [30]. Because myCAFs share many characteristics with myofibroblasts, it is reasonable that CKAP4 activates myCAFs similarly to myofibroblasts, promotes an aggressive phenotype of BCa and worsens survival outcomes [31,32]. However, CKAP4 expression in CAFs was not significantly associated with poor CSS or RFS. In terms of RFS, the CKAP4 positive group was significantly associated with cases that received prior AC, suggesting that CKAP4 expression did not induce poor RFS due to the treatment effect of AC. In contrast, CKAP4 positivity was not associated with SC, and the reason it did not cause significantly poorer CSS remains unclear. Although these findings are intriguing, the functional role of CKAP4 in CAFs was not analyzed in this study. Further investigations, including in vitro and in vivo studies, are needed to elucidate its biological significance.

To our knowledge, this is the first clinical study to demonstrate that CKAP4 expression in tumor tissue may be a useful prognostic biomarker for patients with BCa who undergo RC. Unlike currently available diagnostic biomarkers such as NMP22 and BTA, CKAP4 provided a novel insight into the risk of postoperative recurrence. Moreover, recent studies have reported serum or urine-based biomarkers [33,34]. In the present study, CKAP4 was evaluated in surgical specimens, and its expression can be considered to be independent of serum or urinary biomarkers. This suggests that CKAP4 could be used in combination with serum or urinary biomarkers to improve clinical decision-making.

While searching for literature on CKAP4 in PubMed and other databases, reports related to p63 were often presented in the results. Although both gene products have a molecular weight of 63,000, the CKAP4 protein is mainly localized in the cytoplasm and cell membrane [12,17,35], whereas the p63 protein is expressed in the nucleus and is known as one of the p53-related antigens [36]. Furthermore, the CKAP4 gene is located on chromosome 12q23.3 (OMIM: 618595), while the p63 gene is on chromosome 3q27-28 (OMIM: 603273). Therefore, these are different molecules. The expression of CKAP4 in human normal tissues and various tumors, as reported by The Human Protein Atlas (http://www.proteinatlas.org), is also localized in the cytoplasm and cell membrane; therefore, these proteins should not be confused with one another.

The present study had some limitations. First, it was a single-center, retrospective study with a limited sample size. Second, RC was performed by different surgeons, and the selection of postoperative chemotherapy regimens such as AC and SC depended on the preference of each doctor; this may have affected the results. Third, we did not include patients who received immune-checkpoint inhibitors or enfortumab vedotin. Future studies including these therapies may show different results. Fourth, some patient characteristics such as smoking status and occupational exposure to industrial chemicals or toxins were not included, although they may have influenced patients' prognosis. Further studies addressing these limitations are required to validate our findings. Finally, although CKAP4 expression in CAFs was evaluated, its biological function was not investigated in this study. The functional significance should be explored using in vitro and in vivo models in future studies.

## 5. Conclusions

The expression CKAP4 in tumor cells and CAFs was significantly associated with an aggressive BCa phenotype. CKAP4 expression in tumors was an independent prognostic factor for RFS in patients with BCa who underwent RC. The findings of this study indicate that CKAP4 expression in tumors has potential as a novel biomarker for predicting tumor aggressiveness and poor prognosis in cases of BCa. Previous studies on CKAP4 in BCa are limited and involved basic research. Therefore, this clinical study using real-world data is crucial for determining the usefulness of CKAP4 as a novel clinical biomarker for BCa.

**Author Contributions:** Conceptualization, H.K., Y.S. (Yuichi Sato) and K.M.; methodology, A.S., K.Y., Y.S. (Yuriko Shimizu), Y.S. (Yuichi Sato) and K.M.; formal analysis, H.K., D.K. and Y.S. (Yuichi Sato); investigation, H.K., D.K., S.H., Y.S. (Yuriko Shimizu), W.H., S.S., M.I., H.T. and D.I.; data curation, H.K.; writing—original draft preparation, H.K.; writing—review and editing, D.K., Y.S. (Yuichi Sato) and K.M.; supervision, K.M. All authors have read and agreed to the published version of the manuscript.

**Funding:** This study was supported in part by the Japan Society for the Promotion of Science Grants-in-Aid for Scientific Research (KAKENHI; Grant No. JP24K12515 to K.M.).

**Institutional Review Board Statement:** This study was conducted in accordance with the guidelines of the Declaration of Helsinki and was approved by the Institutional Review Board of Kitasato University School of Medicine and Hospital on 24 May 2017 (Protocol code B17-010).

**Informed Consent Statement:** Informed consent was not individually obtained from all participants, as this study was conducted retrospectively. Information regarding the study and the opportunity to opt-out was provided through our institutional website and posters.

**Data Availability Statement:** The datasets used and/or analyzed during the study are available from the corresponding author upon reasonable request.

**Conflicts of Interest:** Author Yuichi Sato was employed by the company KITASATO-OTSUKA Biomedical Assay Laboratories Co., Ltd. The remaining authors declare that the research was conducted in the absence of any commercial or financial relationships that could be construed as a potential conflict of interest.

## Abbreviations

The following abbreviations are used in this manuscript:

| | |
|---|---|
| CKAP4 | Cytoskeleton-associated protein 4 |
| BCa | Bladder cancer |
| RC | Radical cystectomy |
| RFS | Recurrence-free survival |
| LN | Lymph node |
| MIBC | Muscle-invasive bladder cancer |
| NMIBC | Non-muscle-invasive bladder cancer |
| NAC | Neoadjuvant chemotherapy |
| AC | Adjuvant chemotherapy |
| LVI | Lymphovascular invasion |
| CAF | Cancer-associated fibroblast |
| CIS | Carcinoma in situ |
| SC | Salvage chemotherapy |
| TBS | Tris-buffered saline |
| CSS | Cancer-specific survival |
| myCAF | Myofibroblast-like cancer-associated fibroblast |

## References


1. Sung, H.; Ferlay, J.; Siegel, R.L.; Laversanne, M.; Soerjomataram, I.; Jemal, A.; Bray, F. Global cancer statistics 2020: GLOBOCAN estimates of incidence and mortality worldwide for 36 cancers in 185 countries. *CA Cancer J. Clin.* **2021**, *71*, 209–249. [CrossRef] [PubMed]
2. Hautmann, R.E.; de Petriconi, R.C.; Pfeiffer, C.; Volkmer, B.G. Radical cystectomy for urothelial carcinoma of the bladder without neoadjuvant or adjuvant therapy: Long-term results in 1100 patients. *Eur. Urol.* **2012**, *61*, 1039–1047. [CrossRef]
3. Witjes, J.A.; Compérat, E.; Cowan, N.C.; De Santis, M.; Gakis, G.; Lebret, T.; Ribal, M.J.; Van der Heijden, A.G.; Sherif, A. EAU guidelines on muscle-invasive and metastatic bladder cancer. *Eur. Urol.* **2014**, *65*, 778–792. [CrossRef] [PubMed]
4. Shariat, S.F.; Tokunaga, H.; Zhou, J.H.; Kim, J.H.; Ayala, G.E.; Benedict, W.F.; Lerner, S.P. p53, p21, pRB, and p16 expression predict clinical outcome in cystectomy with bladder cancer. *J. Clin. Oncol.* **2004**, *22*, 1014–1024. [CrossRef]
5. Shariat, S.F.; Bolenz, C.; Godoy, G.; Fradet, Y.; Ashfaq, R.; Karakiewicz, P.I.; Isbarn, H.; Jeldres, C.; Rigaud, J.; Sagalowsky, A.I.; et al. Predictive value of combined immunohistochemical markers in patients with pT1 urothelial carcinoma at radical cystectomy. *J. Urol.* **2009**, *182*, 78–84; discussion 84. [CrossRef] [PubMed]
6. Lotan, Y.; Bagrodia, A.; Passoni, N.; Rachakonda, V.; Kapur, P.; Arriaga, Y.; Bolenz, C.; Margulis, V.; Raj, G.V.; Sagalowsky, A.I.; et al. Prospective evaluation of a molecular marker panel for prediction of recurrence and cancer-specific survival after radical cystectomy. *Eur. Urol.* **2013**, *64*, 465–471. [CrossRef]
7. Schweizer, A.; Ericsson, M.; Bächi, T.; Griffiths, G.; Hauri, H.P. Characterization of a novel 63 kDa membrane protein. Implications for the organization of the ER-to-Golgi pathway. *J. Cell Sci.* **1993**, *104*, 671–683. [CrossRef]
8. Li, S.X.; Li, J.; Dong, L.W.; Guo, Z.Y. Cytoskeleton-associated protein 4, a promising biomarker for tumor diagnosis and therapy. *Front. Mol. Biosci.* **2020**, *7*, 552056. [CrossRef]
9. Schweizer, A.; Rohrer, J.; Hauri, H.P.; Kornfeld, S. Retention of p63 in an ER-Golgi intermediate compartment depends on the presence of all three of its domains and on its ability to form oligomers. *J. Cell Biol.* **1994**, *126*, 25–39. [CrossRef]
10. Chen, Z.Y.; Wang, T.; Gan, X.; Chen, S.H.; He, Y.T.; Wang, Y.Q.; Zhang, K.H. Cytoskeleton-associated membrane protein 4 is upregulated in tumor tissues and is associated with clinicopathological characteristics and prognosis in hepatocellular carcinoma. *Oncol. Lett.* **2020**, *19*, 3889–3898. [CrossRef]
11. Kajiwara, C.; Fumoto, K.; Kimura, H.; Nojima, S.; Asano, K.; Odagiri, K.; Yamasaki, M.; Hikita, H.; Takehara, T.; Doki, Y.; et al. p63-dependent Dickkopf3 expression promotes esophageal cancer cell proliferation via CKAP4. *Cancer Res.* **2018**, *78*, 6107–6120. [CrossRef] [PubMed]
12. Sun, C.M.; Geng, J.; Yan, Y.; Yao, X.; Liu, M. Overexpression of CKAP4 is associated with poor prognosis in clear cell renal cell carcinoma and functions via cyclin B signaling. *J. Cancer* **2017**, *8*, 4018–4026. [CrossRef]
13. Kimura, H.; Yamamoto, H.; Harada, T.; Fumoto, K.; Osugi, Y.; Sada, R.; Maehara, N.; Hikita, H.; Mori, S.; Eguchi, H.; et al. CKAP4, a DKK1 receptor, is a biomarker in exosomes derived from pancreatic cancer and a molecular target for therapy. *Clin. Cancer Res.* **2019**, *25*, 1936–1947. [CrossRef]
14. Sun, X.; Xie, L.; Qiu, S.; Li, H.; Zhou, Y.; Zhang, H.; Zhang, Y.; Zhang, L.; Xie, T.; Chen, Y. Elucidation of CKAP4-remodeled cell mechanics in driving metastasis of bladder cancer through aptamer-based target discovery. *Proc. Natl. Acad. Sci. USA* **2022**, *119*, e2110500119. [CrossRef] [PubMed]
15. Matsumoto, K.; Tabata, K.; Hirayama, T.; Shimura, S.; Nishi, M.; Ishii, D.; Fujita, T.; Iwamura, M. Robot-assisted laparoscopic radical cystectomy is a safe and effective procedure for patients with bladder cancer compared to laparoscopic and open surgery: Perioperative outcomes of a single-center experience. *Asian J. Surg.* **2019**, *42*, 189–196. [CrossRef]
16. Matsumoto, K.; Ikeda, M.; Sato, Y.; Kuruma, H.; Kamata, Y.; Nishimori, T.; Tomonaga, T.; Nomura, F.; Egawa, S.; Iwamura, M. Loss of periplakin expression is associated with pathological stage and cancer-specific survival in patients with urothelial carcinoma of the urinary bladder. *Biomed. Res.* **2014**, *35*, 201–206. [CrossRef] [PubMed]
17. Yanagita, K.; Nagashio, R.; Jiang, S.X.; Kuchitsu, Y.; Hachimura, K.; Ichinoe, M.; Igawa, S.; Fukuda, E.; Goshima, N.; Satoh, Y.; et al. Cytoskeleton-associated protein 4 is a novel serodiagnostic marker for lung cancer. *Am. J. Pathol.* **2018**, *188*, 1328–1333. [CrossRef]
18. Nagoya, A.; Sada, R.; Kimura, H.; Yamamoto, H.; Morishita, K.; Miyoshi, E.; Morii, E.; Shintani, Y.; Kikuchi, A. CKAP4 is a potential exosomal biomarker and therapeutic target for lung cancer. *Transl. Lung Cancer Res.* **2023**, *12*, 408–426. [CrossRef]
19. Akanda, M.R.; Ahn, E.J.; Kim, Y.J.; Salam, S.M.A.; Noh, M.G.; Kim, S.S.; Jung, T.Y.; Kim, I.Y.; Kim, C.H.; Lee, K.H.; et al. Different Expression and Clinical Implications of Cancer-Associated Fibroblast (CAF) Markers in Brain Metastases. *J. Cancer* **2023**, *14*, 464–479. [CrossRef]
20. Saito, D.; Tadokoro, R.; Nagasaka, A.; Yoshino, D.; Teramoto, T.; Mizumoto, K.; Funamoto, K.; Kidokoro, H.; Miyata, T.; Tamura, K.; et al. Stiffness of primordial germ cells is required for their extravasation in avian embryos. *IScience* **2022**, *25*, 105629. [CrossRef]
21. Innocenti, M. New insights into the formation and the function of lamellipodia and ruffles in mesenchymal cell migration. *Cell Adh. Migr.* **2018**, *12*, 401–416. [CrossRef] [PubMed]

22. Carmona, G.; Perera, U.; Gillett, C.; Naba, A.; Law, A.L.; Sharma, V.P.; Wang, J.; Wyckoff, J.; Balsamo, M.; Mosis, F.; et al. Lamellipodin promotes invasive 3D cancer cell migration via regulated interactions with Ena/VASP and SCAR/WAVE. *Oncogene* **2016**, *35*, 5155–5169. [CrossRef]
23. Giralt, I.; Gallo-Oller, G.; Navarro, N.; Zarzosa, P.; Pons, G.; Magdaleno, A.; Segura, M.F.; Sábado, C.; Hladun, R.; Arango, D.; et al. Dickkopf-1 inhibition reactivates Wnt/β-catenin signaling in rhabdomyosarcoma, induces myogenic markers in vitro and impairs tumor cell survival in vivo. *Int. J. Mol. Sci.* **2021**, *22*, 12921. [CrossRef] [PubMed]
24. Niida, A.; Hiroko, T.; Kasai, M.; Furukawa, Y.; Nakamura, Y.; Suzuki, Y.; Sugano, S.; Akiyama, T. DKK1, a negative regulator of Wnt signaling, is a target of the β-catenin/TCF pathway. *Oncogene* **2004**, *23*, 8520–8526. [CrossRef]
25. Shinno, N.; Kimura, H.; Sada, R.; Takiguchi, S.; Mori, M.; Fumoto, K.; Doki, Y.; Kikuchi, A. Activation of the Dick-kopf1-CKAP4 pathway is associated with poor prognosis of esophageal cancer and anti-CKAP4 antibody may be a new therapeutic drug. *Oncogene* **2018**, *37*, 3471–3484. [CrossRef] [PubMed]
26. Sada, R.; Yamamoto, H.; Matsumoto, S.; Harada, A.; Kikuchi, A. Newly developed humanized anti-CKAP4 antibody suppresses pancreatic cancer growth by inhibiting DKK1-CKAP4 signaling. *Cancer Sci.* **2024**, *115*, 3358–3369. [CrossRef] [PubMed]
27. Shen, C.H.; Hsieh, H.Y.; Wang, Y.H.; Chen, S.Y.; Tung, C.L.; Wu, J.D.; Lin, C.T.; Chan, M.W.Y.; Hsu, C.D.; Chang, D. High Dickkopf-1 expression is associated with poor prognosis in patients with advanced urothelial carcinoma. *Exp. Ther. Med.* **2010**, *1*, 893–898. [CrossRef]
28. Gladka, M.M.; Molenaar, B.; de Ruiter, H.; van der Elst, S.; Tsui, H.; Versteeg, D.; Lacraz, G.P.A.; Huibers, M.M.H.; van Oudenaarden, A.; van Rooij, E. Single-cell sequencing of the healthy and diseased heart reveals cytoskeleton-associated protein 4 as a new modulater of fibroblasts activation. *Circulation* **2018**, *138*, 166–180. [CrossRef]
29. Yang, D.; Liu, J.; Qian, H.; Zhuang, Q. Cancer-associated fibroblasts: From basic science to anticancer therapy. *Exp. Mol. Med.* **2023**, *55*, 1322–1332. [CrossRef]
30. Du, Y.H.; Sui, Y.Q.; Cao, J.; Jiang, X.; Wang, Y.; Yu, J.; Wang, B.; Wang, X.Z.; Xue, B.X. Dynamic changes in myofibroblasts affect the carcinogenesis and prognosis of bladder cancer associated with tumor microenvironment remodeling. *Front. Cell Dev. Biol.* **2022**, *10*, 833578. [CrossRef]
31. Thinyakul, C.; Sakamoto, Y.; Shimoda, M.; Liu, Y.; Thongchot, S.; Reda, O.; Nita, A.; Sakamula, R.; Sampattavanich, S.; Maeda, A.; et al. Hippo pathway in cancer cells induces NCAM1+αSMA+ fibroblasts to modulate tumor microenvironment. *Commun. Biol.* **2024**, *7*, 1343. [CrossRef] [PubMed]
32. Lavie, D.; Ben-Shmuel, A.; Erez, N.; Scherz-Shouval, R. Cancer-associated fibroblasts in the single-cell era. *Nat. Cancer* **2022**, *3*, 793–807. [CrossRef] [PubMed]
33. Ofner, H.; Laukhtina, E.; Hassler, M.R.; Shariat, S.F. Blood-Based Biomarkers as Prognostic Factors of Recurrent Disease after Radical Cystectomy: A Systematic Review and Meta-Analysis. *Int. J. Mol. Sci.* **2023**, *24*, 5846. [CrossRef] [PubMed]
34. Malinaric, R.; Mantica, G.; Monaco, L.L.; Mariano, F.; Leonardi, R.; Simonato, A.; der Merwe, A.V.; Terrone, C. The Role of Novel Bladder Cancer Diagnostic and Surveillance Biomarkers-What Should a Urologist Really Know? *Int. J. Environ. Res. Public Health* **2022**, *19*, 9648. [CrossRef]
35. Tuffy, K.M.; Planey, S.L. Cytoskeleton-associated protein 4: Functions beyond the endoplasmic reticulum in physiology and disease. *ISRN Cell Biol.* **2012**, *2012*, 142313. [CrossRef]
36. Di Como, C.J.; Urist, M.J.; Babayan, I.; Drobnjak, M.; Hedvat, C.V.; Teruya-Feldstein, J.; Pohar, K.; Hoos, A.; Cordon-Cardo, C. p63 expression profiles in human normal and tumor tissues. *Clin. Cancer Res.* **2002**, *8*, 494–501. [CrossRef]

MDPI AG
Grosspeteranlage 5
4052 Basel
Switzerland
Tel.: +41 61 683 77 34

*Cancers* Editorial Office
E-mail: cancers@mdpi.com
www.mdpi.com/journal/cancers